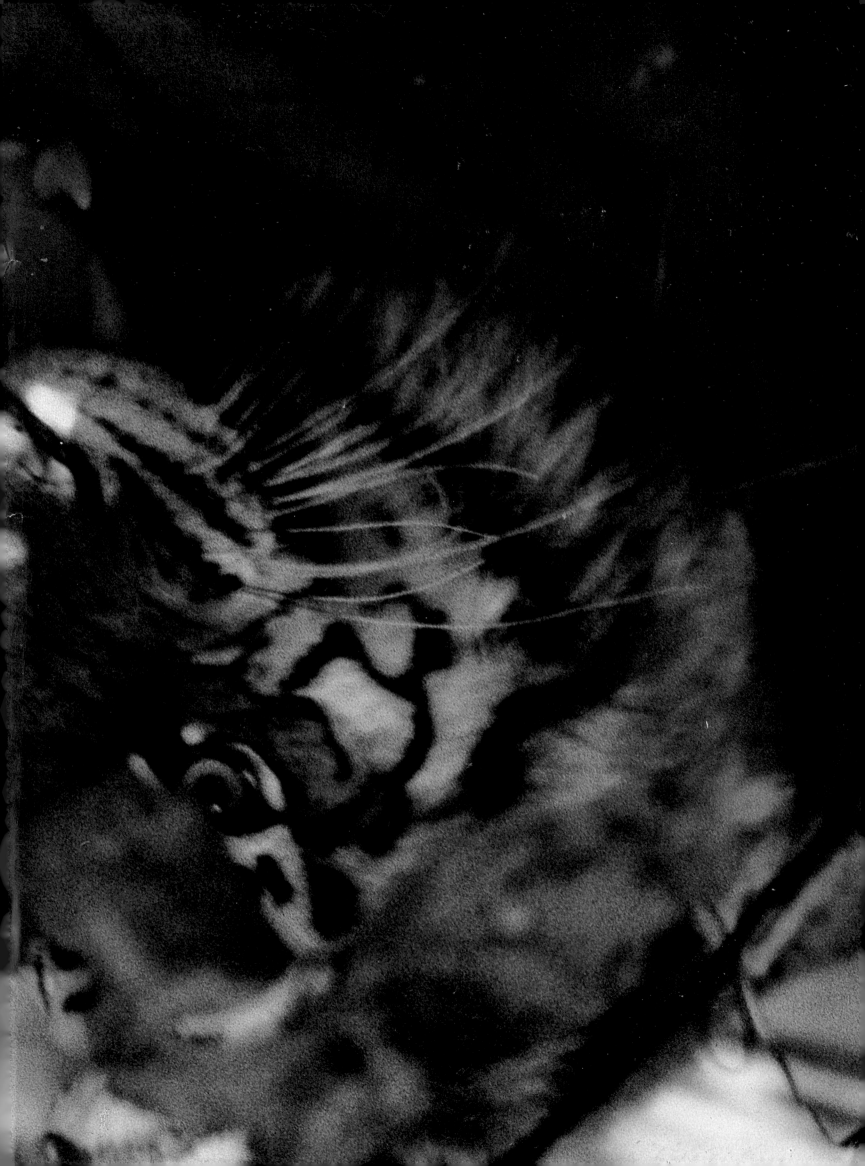

THE AUDUBON SOCIETY BOOK OF WILD ANIMALS

Les Line, Editor of Audubon magazine, and Edward Ricciuti

THE AUDUBON SOCIETY

A Chanticleer Press Edition

BOOK OF WILD ANIMALS

HARRY N. ABRAMS, INC., PUBLISHERS, NEW YORK

Library of Congress Catalog Card Number: 77-9159
Line, Les.
The Audubon Society Book of Wild Animals.
Includes index.
1. Mammals. I. Ricciuti, Edward R., joint author.
II. National Audubon Society. III. Title.
QL706.L56   599   77-9159
ISBN 0-8109-0670-8
Second Printing

Prepared and produced by Chanticleer Press, Inc.
Color reproductions by Cliché & Litho AG, Zurich, Switzerland
Printed and bound by Amilcare Pizzi, S.p.A., Milan, Italy

Chanticleer Staff:
Publisher: Paul Steiner
Editor-in-Chief: Milton Rugoff
Managing Editor: Gudrun Buettner
Project Editor: Susan Rayfield
Assistant Editor: Liz Kaufmann
Production: Emma Staffelbach
Design: Massimo Vignelli

*Note on Illustration Numbers: All illustrations are numbered
according to the pages on which they appear.*

First frontispiece: *A tiger* (Panthera tigris) *peering from
an Indian forest.* (Belinda Wright)
Second frontispiece: *A young male leopard* (Panthera pardus)
in the jungle of Sri Lanka. (George Holton)
Title page: *A yawning lion* (Panthera leo) *in Tanzania's Ngorongoro Crater.*
(M. Philip Kahl, Jr.)
Third frontispiece: *Cheetahs* (Acinonyx jubatus) *surrounding a kill on the
Serengeti Plains.* (Russ Kinne/Photo Researchers)

Contents

Foreword   13
Introduction   15

Life in the Trees   21
Marsupial Marvels   53
Mammals with Wings   69
Gnawing Hordes   85
The Weasel Clan   101
Ancient and Unusual   113
The Hunters: Dogs   125
The Hunters: Cats   141
Big Bears and their Kin   161
The Deer Tribe   173
Grazing Herds   193
The Domesticated Ones   217
Unique Giants   233
Life on the Peaks   245
Mammals in the Sea   261

Notes on Photographers   289
Index   293

For Charles H. Callison, spokesman for the Audubon cause

Foreword

This book represents, of course, only a sampling of the
mammal kingdom—the 181 photographs in it portray
scarcely a twentieth of the world's four thousand
species of mammals. But it is, I think the reader will
agree, a superb sampling, and an appropriate companion
to *The Audubon Society Book of Wild Birds*.
Like its predecessor, it required a team effort to bring
it to fruition. Susan Rayfield, *Audubon* magazine
picture editor, assembled thousands of mammal
photographs from every niche in the world, screened
out the best several hundred, and she and I over
several long winter nights argued the final selection.
(Enough great pictures were left over to fill
another volume of equal beauty.)
Edward R. Ricciuti, knowledgeable and literate in
many fields of natural history and a frequent contributor
to the pages of *Audubon*, shared the authorship; his
informative essays introduce each chapter, while my
captions accompany the individual photographs.
Production was again in the expert hands of the staff of
Chanticleer Press, but in particular this book owes
its concept to Paul Steiner, who believed with fervor
that the world of mammals could be mined for a book
every bit as spectacular as the one on birds. And he
was correct, as the evidence in hand proves.

Les Line

Introduction

"Free as a mammal." Somehow it doesn't sound right.
Yet that statement is no less correct than "Free as
a bird." Therein lies an irony.

The passions and sentiments that mammals arouse in
man bridge the gamut of human emotions—affection,
love, envy, awe, fear, hate. But these feelings are almost
entirely in response to particular kinds of mammals
—dogs, for instance, or bears—or even to individuals.
Not to mammals as a group.

We talk collectively about birds as though there is no
difference between an ostrich, a penguin, and a
hummingbird. It is "the birds" that sing, and "the birds"
that build nests. Whether a person goes afield to
observe warblers, hawks, or geese, he or she is
a "bird watcher." The term is in the dictionary.

But try looking for "mammal watcher." There simply is
no word for the person who walks to the meadow in
the evening to wait for deer, or enjoys the antics of
squirrels in the park, or follows the trail of a hunting
fox in the snow.

Nor do mammals enjoy the broad popular appeal of
birds. Nearly every town of at least modest size has its
local bird club. Great organizations like the National
Audubon Society, the Royal Society for the Protection
of Birds, and the International Council for Bird
Preservation owe their birth to the delight people
share in birds. There are a few professional societies
for scientists specializing in the study of mammals, but
nothing more. No immense, organized following.

Feeding birds has become a phenomenon of our times.
Designers labor over new devices to dispense the

millions of tons of bird seed sold annually. But few people feed wild mammals, except casually or unintentionally—the raccoon that comes to the back door for a handout, or the squirrel that commandeers the bird feeder.

Field guides to the birds are abundant, field guides to the mammals few indeed. Bird watchers go to great extremes and expense to add new species to their "life lists." Few nature enthusiasts keep track of the different mammals they have seen and identified. Birds are the subject of innumerable paeans and poems, mammals of a handful. Few mammals inspire the superlatives we accord the passing of a flight of swans, the dive of a hunting falcon, the song of a thrush.

If any single point of view dominates our attitude toward mammals, it is that their worth is determined by their utility. And this has not worked out particularly well for the mammals. There is nothing intrinsically wrong with using what nature provides. The problem arises when reasonable or necessary use becomes exploitation, when greed jeopardizes survival.

Some birds, of course, are hunted for sport or food. A few have been domesticated. Others are caged for the beauty of their form, color, or song. It also is true that man is responsible for the extinction of the passenger pigeon, the Carolina parakeet, the great auk, and others, either deliberately or indirectly through the destruction of habitat. And near the turn of this century, the plumed birds were nearly wiped out to provide feathers for women's hats.

By and large, however, our attitude toward birdlife has been sympathetic, if not reverent.

The mammals have been less fortunate. They are valued not for what they are, but for what they can do for man. Can they be ridden, harnessed, petted, eaten, or are their skins or tusks precious? Worse, many mammals are considered a threat to our prosperity. Although the Indians of the North American plains prized and revered the buffalo, to the settlers who poured into the West after the Civil War the great shaggy beast was an enemy. Not only did it take rangeland needed for livestock or crops, it provided food and comfort to the Indian foe. And, of course, its hide and meat were marketable. Thus sixty million buffalo were slaughtered in a few years; often only the tongue was salvaged. In 1875, the Texas legislature met at Austin to discuss a bill for the protection

of the fast-shrinking buffalo herds. General Phil
Sheridan, charged by the U.S. Army with subjugating
the southwestern Indians, testified against the
proposal:
"Let them kill, skin, and sell until the buffalo is
exterminated, as it is the only way to bring about
a lasting peace and allow civilization to advance."
Dozens of other mammals have been exploited just as
destructively. Millions of fur seals, sea otters, and
beavers were killed for their pelts with little concern
for the future. The great herds of whales have been
reduced to pitiful remnants. Several species of
dolphins may be exterminated by tuna fishermen who
drown them in their nets, after the boats are directed
to the schools of tuna by the dolphins. There is grave
fear that ivory poaching soon will doom the African
elephant. The luxury fur trade was responsible
for the near demise of many of the spotted cats of
Asia and Africa. In another kind of exploitation, but
equally devastating, the populations of many primates,
like the orangutan, have been reduced to dangerously
low levels by the capture of animals for the pet,
zoo, and laboratory trade.
And other mammals have gotten in man's way—
predators like the wolf, coyote, and cougar, hunted
down and trapped and poisoned because they
occasionally eat livestock. Or prairie dogs, poisoned
by the millions because they compete for the sparse
grasslands claimed for cattle, and as the "dogtowns"
vanished so too did the black-footed ferret, which
preys on the rodents.
However, this does not explain entirely why we view
mammals so restrictively, as compared with birds. Is
it because birds are beautiful, emblazoned with
beautiful colors, and so graceful? Is it because their
songs herald spring and have inspired great composers
to write lovely music? Is it because we are envious
of their power of flight, their ability to vanish into
the clouds, to soar as if they had no link with the earth,
to hurtle from the heavens with incredible speed?
But pause and consider, please. Is the plumage of any
bird more brilliant than the blue and scarlet on the
face of a mandrill? Lovelier than the amber, black,
and white that stamp the tiger? Is any bird more
graceful than a racing dolphin?
Song? From podiums high in the rain forests of Southeast
Asia, the gibbons make music that is as haunting as

any sounds in nature. Its magic was described in *Wild Heritage* by Sally Carrighar, who followed the gibbon song as it rose "by halftone steps to exactly the height of an octave, where the voices trilled with great flexibility. Each tone was introduced by a grace note, the keynote E, and the whole up-flung roulade had the effect of expressing triumphant joy. At the top, during the trill, the gibbons quivered through all their bodies. Finally they let the song ease away in a few diminishing quarter-notes."

No less stirring is the cry of a wolf echoing across the Alaskan tundra, or the bugling of an elk on a chilly autumn evening in a Rocky Mountain meadow.

Flight? More than any other attribute, this is the one most closely associated with birds—although not all birds fly, and many species fly with merely modest skills. However, several mammals—flying squirrels and the marsupial opossums of Australia—engage in swooping glides which are spectacular examples of unpowered flight. And one out of seven of all living species of mammals—the vast order of bats—is capable of powered flight with speed and maneuverability to equal nearly any bird.

In truth, the answer to the great appeal of birds may lie in the fact that they are so evident, while the human experience with mammals is limited. Most birds are active by day, and one can see birds in considerable numbers even in the heart of a city. Mammals, in contrast, are generally nocturnal or crepuscular and retreat into concealment when the sun rises. Most of them shun people and civilization. Even when they coexist in close proximity with man, they are adept at staying out of sight of their neighbors. John Kieran, in his book *Natural History of New York City*, describes how even a great metropolis can harbor myriad mammals whose presence is hardly suspected:

"As for 'wild animals' within the city limits," Kieran writes, "New York has a far larger population of wild quadrupeds than most of the resident bipeds suspect. That's because so many of the wild mammals are small in size and nocturnal in habit, as a result of which they are rarely seen except by those who look for them."

Of course, a few mammals can be observed in multitudes by day—the fur seals which gather in immense colonies, or herbivores such as the wildebeests which form large herds on the African plains. But they are restricted

to dwindling havens far from the masses of humanity.
The mammals deserve to be celebrated and treasured for
all their magic and mystery, for their fascinating
modes of living, and for the curiosities among them.
This book provides some insights into their life and
lore, and examines certain unusual facts about them,
but it is not a treatise on biology. Rather, it is a
tribute to the mammals, applauding them for their
wonder and their accomplishments. It is our means,
through the greatest photographs of mammals to be
found in the world, of awakening greater public interest
in their ways and, ultimately, in their survival.

Edward Ricciuti

Life in the Trees

From the ground, the animals that teem in the forest
canopy of Southeast Asia are largely invisible, but a
multitude of sounds reveals their presence in the tangle
of leaves and lianas. Myriad small creatures chitter and
chatter. Cicadas hum unmistakably, like miniature buzz
saws. Now and then the canopy trembles in the small
windstorm set in motion by the mighty wings of the
great hornbill as it moves to a new perch. And in the
cool of the early morning, as well as the shadowed
hours of the late afternoon, the gibbons begin to call,
often from perches in the loftiest crowns of trees. Their
melodious whoops ring through the green galleries,
echoing from the arboreal world above, a haunting
reminder that all of the primate tribe, even those which
walk with two legs on the ground, have been shaped
and indelibly stamped by life in the trees.
Developing from primitive, shrewlike creatures while
the dinosaurs still walked the land, the primates quickly
found their niche, and their destiny, in the branches.
Man and a few large monkeys such as the mandrill *(Mandrillus sphinx)* and the gorilla *(Gorilla gorilla)* have
returned for the most part to the ground, but there are
times when even they show an affinity for the trees.
When darkness falls, all but the large heavy male
gorillas often will construct a nest of vegetation in the
branches and go to sleep in it. For most primates,
however, the trees are not merely a refuge but home and
feeding grounds. Adaptation to arboreal living has been
most refined in the gibbons *(Hylobates)* and the closely
related siamang *(Symphalangus syndactylus)*, most agile
of all mammals in the boughs. These graceful apes

venture down to the ground only rarely, seldom by choice, and are almost awkward there. In the canopy, they fling themselves into the void between branches with hair-raising abandon, arresting their flight by hooking elongate hands around limb or liana, then hurling themselves into space again.

The ability to wrap a hand around a limb and grasp it — to hold on to something — is a superb advantage for a creature living in the trees, and began to appear almost at the outset of primate existence. Claws slowly changed to flattened nails and long, sensitive digits. Eventually, as a thumb opposable to the other digits developed, the primate hand became capable of such delicate maneuvers as picking up a twig and probing for insects within a termite mound, something that chimpanzees *(Pan troglodytes)* accomplish with ease. Using hands and fingers — and feet and toes — in the branches was demanding to the visual organs.

Gradually, over the course of evolution, the eyes moved from the sides of the head to the front, providing better depth perception and eventually the ability to see the world in three dimensions. Meanwhile, the cerebral covering expanded to coordinate the flood of nerve messages arising from increasingly complex inter-workings of hand and eye. The gray cortex of the brain not only grew in bulk but gained more surface area for multiplying cells by wrinkling until it was covered with folds and convolutions.

The evolution of primate history over more than 70 million years is reflected — although by no means exactly — by the primates that roam the tropics and their margins today. Primate origins among the insectivores are vividly demonstrated by the most primitive of the tribe, the active, aggressive tree shrews (Tupaiidae) of Southeast Asia. Clawed, long of snout, and often bushy-tailed, the tree shrews straddle the taxonomic fence between the primates and the insectivores, and in their general aspect and behavior hint as well at the probable relationships between the very early primates and rodents. By 50 million years ago, primates of a more advanced type than tree shrews flourished over a vast portion of the earth, in forests that covered much of Europe and North America. These were the prosimians, whose eyes faced forward, or nearly so, and whose hands and feet bore slender, flexible digits. Long since vanished from the temperate regions, the prosimians survive today in the jungles. Madagascar is the home of the aye-aye *(Daubentonia madagascariensis)*, the lemurs (Lemuridae), and the indris (Indridae). The lorises and pottos (Lorisidae) inhabit Asia and Africa, respectively. Galagos (Galagidae) are also African, whereas the diminutive, saucer-eyed tarsiers (Tarsiidae) haunt the forests of the Philippines and the Indonesian region. Today the prosimians are in eclipse, restricted to dark places or to isolated backwaters. Their state reflects what happened to them about 30 million years ago, when the monkeys (Cebidae, Callithricidae, Cercopithecidae) and apes (Pongidae) became rulers of the trees.

To watch a South American squirrel monkey *(Saimiri)* manipulate a grape, turning it round and round in its fingers, cocking its head and following the soft little sphere with its eyes, is to know how far the monkeys and apes have come from the tree shrews. Binocular color vision makes the world a different place for them, and fingers capable of fine, independent movement, together with a relatively gigantic brain, permit exploration of that world. Seeing an African talapoin monkey *(Miopithecus talapoin)* run along a limb, secure in a world of swaying uncertainty, is an awakening. Primates have the sophisticated mental equipment they need to react in a flash to the unexpected, and to make complex decisions in situations that can shift in a twinkling. A branch, apparently solid, breaks on contact. A predator appears from nowhere among the leaves. The wind blows the landing site out of reach. For millions of years the primates have met such challenges and surmounted them.

When the graceful, long-limbed spider monkey *(Ateles)* dangles from a liana in the South American forest, its sensitive prehensile tail serving as a fifth hand, it testifies along with the gibbons that the primates have attained nearly absolute mastery over the arboreal environment. When the shaggy orangutan *(Pongo pygmaeus)* hangs by its hands from a branch in a Borneo rain forest, body suspended below, it mirrors a step in the process that, once some of the apes had descended to the ground, led to walking upright. The hands that cling to branches might just as well pluck food, or even carry it overland, as chimpanzees do today. The descent to the ground, and the first exploration by apes of the savanna, is an epic forever lost; but glimpses of what it must have been like emerge from watching the meanderings of bands of rhesus monkeys

(*Macaca mulata*) over the landscape of southern Asia, and gelada baboons *(Theropithecus gelada)* romping on the rocky heights of Ethiopia.

The baboon troop is a society on the move, not haphazardly, but in precisely organized fashion. Baboon troops are governed by scrupulously observed patterns of complex social behavior, possibly resembling those of other terrestrial primates which ranged the African grasslands millions of years ago as baboons do today. These ancient primates, still dimly perceived, were apes, but on the verge of becoming human. Within the baboon troop, for example, a youngster assumes new responsibilities with age. Baboon society is above all cohesive, a trait very much suited to survival. As long as the troop is together, no single baboon need face alone a threat from an outside enemy. To challenge even the baboon of lowliest rank is to take on the entire troop. Bands of chimpanzees also threaten aggressors in united fashion, although their groups are not as large or quite so organized as those of the baboons. Chimpanzees have been observed working together in another way; they will surround a monkey so that one of their band, a large male, can move in and kill it for food. Most primates, perhaps with the notable exception of the forest-dwelling gorilla, will at least sample bird's eggs, nestlings, insects, and other small prey. Life on the savanna, however, the mode of living that was to produce the first humans, carries primate predation to the ultimate—the stalking and killing of game and the consumption of raw flesh.

25 and **26** overleaf. Unlike other lemurs of the Indian Ocean islands of Madagascar and Comoro, the two sexes of the black lemur (Lemur macaco) are strikingly different: the male is jet black, the female reddish brown with a pale ruff surrounding her face. Indeed, she was once believed to be an entirely different species and was described in literature as the "white-bearded lemur." Also unlike other arboreal lemurs, the black lemur will drop to the ground if pursued, dash through thick cover to another high tree some distance away, then speed to its crown. It is capable of leaps of more than twenty-five feet. And whereas other lemurs have scent glands on their forearms for use in marking their territories, the black lemur lacks these and instead resort to screams that pierce the jungle in the gathering dusk. Careful grooming of their water-resistant fur is important to lemurs; thus they are equipped with a long claw on the second toe of each hind foot, a rough-edged tongue, and lower teeth that project outward to form a comb. Two black lemurs meeting in the forest will engage in mutual grooming.

28 second overleaf. Rocky country is the home of the ring-tailed lemur (Lemur catta), and troops of fifteen or twenty may be seen parading about on all fours, bushy tails held high in the air. Ring-tailed lemurs, which purr and mew like house cats, love to sunbathe with legs and arms spread wide. Living in arid habitat, they quench their thirst with juicy fruits. Sitting on its haunches, holding fruit in its hands, a lemur delicately bites off pieces with its back teeth so the juice runs into its mouth and not on its fur. A male ring-tailed lemur liberally marks its territory with scent glands on its arms and anal area; it will also impregnate its tail with the foul-smelling fluid and wave it about to impress other males.

31. *Largest of its kind—and one of the loveliest, with its long silky pelage and exaggerated mane—is the golden lion marmoset (*Leontideus rosalia*), a rare inhabitant of coastal mountains near Rio de Janeiro. Using its exceptionally long fingers to grab branches, it scrambles across the forest canopy at unbelievable speeds.*

32 *overleaf. Only five or six inches long, plus an eight-inch tail, the pygmy marmoset (*Cebuella pygmaea*) of Amazonia rivals the mouse lemurs of Madagascar for the distinction of being the world's tiniest primate. Pygmy marmosets breed in treetop dens. The twin babies are the size of navy beans, but they are fully furred, their eyes are open, and they have fingers so small one can see them only with a magnifying lens. The babies are carried about on the groin of their father, and he passes them over to the mother only for nursing.*

34 *second overleaf. The titi monkeys (*Callicebus sp.*) are handsome, colorful residents of forests north of the Amazon; a distinctive mark of several of the species of titis is a white or black band across the forehead. Titis, in contrast to most other monkeys, sleep and nest in hollow trees. Before a family group sets out at dawn to feed on insects, snails, lizards, small birds, and green fruit, it announces its intentions to the jungle at large with a deafening chorus of howls.*

30 *top. Squirrel-sized and squirrel-like in their habits, the marmosets and tamarins of the South American rain forests are among the smallest primates. Many species, like the white-eared marmoset (*Callithrix aurita*) of Brazil's southeastern coastal mountains, sport flowing ruffs, manes, beards, plumes, or ear tufts.*

30 *above. The cottonhead tamarin (*Saguinus oedipus*), found on the Caribbean coast of Colombia, has a snow-white mane that it erects when it is excited. With its long legs and extremely long tail, it seems to fly from tree to tree. More so than other tamarins, the cottonhead is carnivorous, preying on mice, birds, and insects.*

36. *Guenons are long-tailed, multi-colored African monkeys found in almost every kind of habitat south of the Sahara, from sea level to high mountains. Their endlessly-varied color arrangements led earlier zoologists to name more than a hundred different species, but the guenons are lumped today into about a dozen species. The white-bearded De Brazza's monkey (*Cercopithecus neglectus), *known from Cameroun to Kenya, is frequently seen running about on the ground, catching the insects that form a major part of its diet.*

36. *The apparent purpose of the enormous nose of the male proboscis monkey (*Nasalis larvatus) *of Borneo's swamp forests is that of a voice amplifier, for it swells up like a balloon when the monkey sounds its nasal honking call. On an old male, this monstrous appendage dangles below the chin and must be pushed aside when the monkey eats.*

36. *The color of the rare golden langur (*Presbytis geei) *of Bhutan changes with the seasons, from a light chestnut in winter months to a creamy white during the summer. Leaves are the principal diet of langurs, the familiar arboreal monkeys of India and Southeast Asia.*

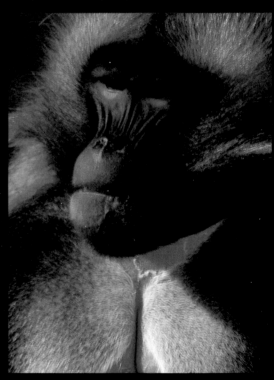

37. *Although it is screaming in rage, crimson is the natural color of the bare face of the red uakari (Cacajao rubicundus). Residents of river-bottom forests in upper Amazonia, uakaris are the only monkeys in the New World that have short tails. Because their habitat is usually flooded, they spend most of their lives in the trees; poor leapers, they travel on all fours along the stout branches of the lower forest canopy.*

37. *Intensifying colors on the red, blue, and purple face of a male mandrill (Mandrillus sphinx) com- municate displeasure to others of lesser rank in a mandrill troop. This largely terrestrial primate of West Africa's equatorial forest can be both a boon and pest to man, depending on whether it is feasting on plagues of locusts or raiding crops.*

37. *When angered or jealous, the gelada (Theropithecus gelada) curls its red lips over its nose and chin, bares huge canine teeth, and flashes pink eyelids, while the bare red patches on its chest deepen in color. This baboon-like monkey of treeless mountain plateaus in northern Ethiopia leads a sedentary life: emerging from caves and crevices in the morning, it will sun for hours before moving a short distance away to feed.*

38. *The booming territorial calls of male howler monkeys—audible for two miles in the jungles of Central and South America—are reputed to be the loudest sounds made by any animal. Bands of as many as forty howlers move about the forests day and night, and their persistent choruses may be heard at any hour, but most often at dawn or dusk. They are the largest monkeys in the New World. Like others of its kind, the red howler monkey (Alouatta* seniculus) *has such a powerful tail that, leaping from a branch, it may stop its flight in midair by not releasing its tail-hold.*

39 *and* **40** *overleaf. Venerated by Hindus, the Indian langur (Presbytis* entellus) *roams villages and even large cities, inhabits temples, raids gardens and crops, and snatches food from tables without fear of harm from man. Thus, in contrast to other langurs, these "civilized" monkeys lead a largely terrestrial life. But totally wild populations of this largest langur retain their arboreal habits, and their leaps of more than thirty feet from branch to branch are nothing short of spec- tacular; sometimes they will change direction in mid-flight. Red hibiscus blossoms are a favored food.*

42-43. *For the first three or four months of its life, an infant Chacma baboon* (Papio ursinus) *hangs upside down beneath its mother's belly, clinging to her breast. But soon it will try to ride on her back, holding tightly with hands and feet until it masters the trick of sitting near her rump, its tiny back braced against the female's stiffly held tail. When she runs, however, the youngster will lie flat. A mother and her newborn infant are the immediate focus of interest in a baboon troop: dominant males hover about and other females groom both the baby and the mother. Old males also play with young baboons, pulling their tails or letting the infants leap on them.*

44-45. *No other monkeys live as far north as the population of Japanese macaques* (Macaca fuscata) *at the tip of the island of Honshu. There, in winter, five feet of snow covers the mountain slopes, forcing the leaf- and fruit-eating monkeys to subsist on tree buds, shoots, and bark. Japanese macaques are highly gregarious, forming societies of as many as 600 monkeys in which the males are organized into a complex class system and even the females have social status. For instance, when foraging macaques stop to feed, they form two circles: the older males join the females, infants, and juveniles in the middle, with the young males making up the outer ring. Strictly protected by the Japanese government, this is the* monkey that symbolizes the wisdom of Buddha: "See no evil, hear no evil, speak no evil."

47. *This young chimpanzee (Pan* troglodytes) *may suckle its mother for the first four years of its life and will remain at her side for several more years. It will not breed until it is at least eleven years old. Believed to be the most intelligent animal next to man, chimps have a vocal repertoire of thirty-two different sounds and are able to use simple tools. Though predominantly vegetarian, they occasionally kill small antelopes and even monkeys.*

46. *The only great ape in Asia, the orangutan* (Pongo pygmaeus) *each night builds a tree platform of sticks and vines for sleeping, covering itself with a blanket of leaves if it is raining. Only a few thousand orangutans survive in Borneo and Sumatra. Uncontrolled traffic in young orangutans for zoos and laboratory experiments depleted the species; now its forests are being cut from beneath it by international timber companies.*

48. *A 400-pound male gorilla (Gorilla gorilla) is six feet tall when it stands upright, its outstretched arms spread eight feet. But it usually prowls the equatorial forests of Africa on all fours, leaning on the knuckles of its partly closed hands. Among man's closest relatives, the gorilla is exclusively vegetarian. At night, females and young in a gorilla band retire to the trees; the gray-haired male leader sleeps in a nest on the ground.*

49 *and* **50** *overleaf. With remarkably long arms, and hands that do not grasp a branch but are used like hooks, the white-handed gibbon* (Hylobates lar) *swings through the rain forests of Malaysia in a blur. Gibbons are the most agile of all primates—indeed, of all mammals— and may cover ten feet in a single swing. Their populations, however, have been decimated by logging of their habitat and by the relentless pet trade in Southeast Asia; young gibbons are captured by killing the mothers.*

Marsupial Marvels

A creature moves in the treetops crowning a ridge in the lush forest of northeastern Australia. Brown of fur, with a yard-long body and an even longer, cylindrical tail, the animal quickly nibbles leaves high above the forest floor. Then, for reasons unknown, it decides it must get to the ground—in a hurry. It launches itself into space, tail streaming behind as a rudder, and sails through the air for almost sixty feet. Once the prodigious leap to the earth has been completed, the creature rises to a half crouch and hops into the shadows, now moving much like any other kangaroo.

The tree kangaroos *(Dendrolagus)* offer dramatic evidence of how the marsupials have diverged as they have adapted to a remarkably wide range of living conditions. In Australia, from which most of the more advanced mammals are absent, the marsupials have taken advantage of the niches left by the lack of competition. And so they have differentiated, while radiating along lines clearly parallel to those taken by the higher placental mammals that have supplanted the marsupials in most parts of the world. Their varying approaches to survival are expressed even in their wide range of body sizes. Consider the gulf between the minuscule planigales *(Planigale)*, which feed on grasshoppers longer than they, and the great gray kangaroo *(Macropus giganteus)*, which is the size of a large man. As a rule, however, marsupials share a basic body plan. In a differing degree, extreme in the kangaroos (Macropodidae), they have longer hind legs than fore-legs. And they have a distinctive taper to their hind parts, particularly where the rump joins the tail. This characteristic is normally associated with reptiles, and is indeed so pronounced in the dog-sized thylacine *(Thylacinus cynocephalus)*, or "marsupial wolf," of Tasmania, that it virtually proclaims the reptilian ancestry of the mammals.

Of course, the external trait most commonly associated with the marsupials—and, in fact, their trademark—is the pouch, although it is missing in a few types. Opening to the front in some, to the rear in others, the pouch is a halfway house between the womb and the outside world. When the marsupial is born it is ill-prepared for life independent of its mother's body. While in the womb, the embryonic marsupial does not have the benefit of the strong placental connection to the uterine wall that nourishes the developing young of more advanced mammals until they are fully formed. The marsupial, therefore, is born before it is fully developed. Unable to see, it still must negotiate its way through its mother's fur to the pouch. That is the most hazardous journey the marsupial ever makes, and not all make it. Once inside, however, the youngster has the shelter, warmth, and food it needs to grow to a fully formed creature.

Beyond its basic characteristics, the marsupial body has considerable range. The peculiarities of each form provide clues to the way of life for which it has evolved. The stocky, badger-like body of the wombat *(Vombatus ursinus)* is built for burrowing. The slender numbat *(Myrmecobius fasciatus)* has a long snout and a ribbon-like tongue that snakes into crevices after termites. The koala *(Phascolarctos cinereus)*, whose rotund body has been immortalized in the form of the teddy bear, has long flexible fingers tipped with hooked claws for life in the trees.

Myriad smaller marsupials, which scurry about the ground like mice and rats, have bodies that outwardly resemble those of such rodents. Several groups of tiny carnivorous marsupials, such as the dunnarts *(Sminthopsis)* and the Kultarrs and wuhl-wuhls *(Antechinus)*, are, in fact, commonly known as "marsupial mice." The pygmy possums and the mundardas *(Cercartetus)* of the phalanger family are similar to dormice. New Guinea, which shares some of the Australian marsupials, has its mouse bandicoot *(Microperoryctes)*, a rare creature known from only three specimens. And, in an illustration of turnabout being fair play, just as there are kangaroo rats among the rodents, the marsupials have rat kangaroos. The smallest, the musky rat kangaroo *(Hypsiprymnodon moschatus)* of Queensland, is ten inches not counting its tail.

The kangaroos show how in Australia, and partly in New Guinea, marsupials have moved into niches occupied in other places by the placental mammals. The great gray kangaroo, the red kangaroo *(Macropus rufus)*, and the euro *(M. robustus)* are browsers and grazers, playing the role that four-footed herbivores assume elsewhere. Each of these large kangaroos specializes in a particular type of habitat. The gray leaps through parklike forests and bush country. The red is a creature of vast inland plains. The euro stays among the rocks, and can survive in sun-baked badlands shunned by its fellow kangaroos. The high stony cliffs are the romping grounds for the spaniel-sized rock wallabies *(Petrogale)*, marsupial counterparts of chamois and mountain goats. Bounding from boulder to boulder, the rock wallabies approach the chamois in their mastery of the heights.

As demonstrated by the tree kangaroos, the marsupials have extended their colonization of habitats into the branches. The cuscuses *(Phalanger)*, saucer-eyed, nocturnal creatures which range as far as Celebes and the Solomon Islands, live entirely in the trees and, like a number of other marsupials, have a prehensile tail. Some of their relatives, such as the greater gliding possum *(Schoinobates volans)* of eastern Australia's high timber, have even evolved flaps of loose skin along the sides of their bodies, so that they can glide, like flying squirrels, from limb to limb.

As predators, marsupials have exploited myriad possibilities, many of which mirror the roles of the true carnivores. The Australian marsupials known as "native cats" *(Dasyurus)*—because to European eyes they resembled felines—fill in for such placental predators as weasels and raccoons. Only two relatively large marsupial predators remain in the Australian region, however, and they are restricted to Tasmania. One, fairly common, is the Tasmanian devil *(Sarcophilus harrisii)*, a feisty beast that is built like a mini-bear, with a head that seems as large as its twenty-pound body and gives the impression of being all muscles and teeth. It might be dubbed the marsupial hyena, because it prowls by night, sometimes making unearthly noises, and scavenges carcasses, using its bone-crushing bite to good advantage. Like hyenas, the devil is not loath to feed upon whatever live prey it can handle, including the young and infirm of larger mammals. Carrion for the devil once was provided in abundance by the thylacine, which, working alone or in pairs, harried kangaroos and wallabies across the open plains and broken forest lands of Tasmania. This dog-sized predator, largest surviving marsupial meat eater, often feeds only on the blood, liver, and kidney fat of its victims, and does not return to a kill. When Europeans arrived with their sheep in Tasmania, and christened the thylacine a "wolf," it behaved as perhaps it was expected to, and raided the flocks. The response from the newcomers was predictable, and today only a handful of thylacines may remain in a few misty mountain fastnesses, hardly more than a memory.

All of the Australian marsupials may have originated from a meat-eating form, which may have resembled the American opossums (Didelphidae). During the final stages of the Age of Reptiles, more than 70 million years ago, opossums were widespread in the world, and could have reached Australia via Africa or Antarctica.

Today, opossums survive only in the Americas, which means the sequence has come full circle for them, because South America probably was the cradle of all the marsupials. The opossums and a virtually unknown family of shrewlike creatures called "rat opossums" (Caenolestidae) are the only survivors of the competition the placental mammals brought to South America about 5 million years ago. But as befits survivors of a difficult battle, they are hardy and adaptable. The Virginia opossum *(Didelphis virginiana)* has even turned the tables by extending its range into the heart of North America. It has pushed as far north as Canada, and the only real problem it has faced is frostbite on its naked ears and tail in very cold weather. It adapts so well to humans that it readily inhabits heavily populated urban areas, even venturing into New York City. The Virginia opossum and its South American relatives are almost exclusively arboreal, and most have prehensile tails. But with typical marsupial diversity, the opossums also have produced a species, the yapok *(Chironectes minimus)*, with webbed feet and a facility in the water that rivals that of the otter. Although the marsupials begin life at a great disadvantage in their competition with the placental mammals, and the domain of the kangaroo and its kin has diminished, the pouched mammals remain testimony to the marvelous flexibility of nature.

57. *The spotted cuscus* (Phalanger maculatus) *is a monkey-like marsupial that inhabits forests from the Cape York Peninsula, Australia, to New Guinea and islands as far as Celebes and the Solomons. The cuscus moves sluggishly and eats quantities of leaves as well as small mammals and birds. Its prehensile tail is furred on top and naked and scaly below, and its tiny ears are hidden beneath woolly fur.*

58 *overleaf. Gliders are the "flying" marsupials of Australia's coastal forests, and the greater glider* (Schoinobates volans) *is, as its name suggests, the largest of all the island continent's airborne possums. It reaches a total length of three feet, including a bushy twenty-inch tail that is used as a rudder during flights that may cover a distance of 250 feet. The gliders have a layer of skin on each side of the body that, when outstretched, becomes their parachute for swooping down from high branches. Like all possums, the mother glider carries her young on her back once they have outgrown the pouch. The big eyes and ears are adaptations for a nocturnal life.*

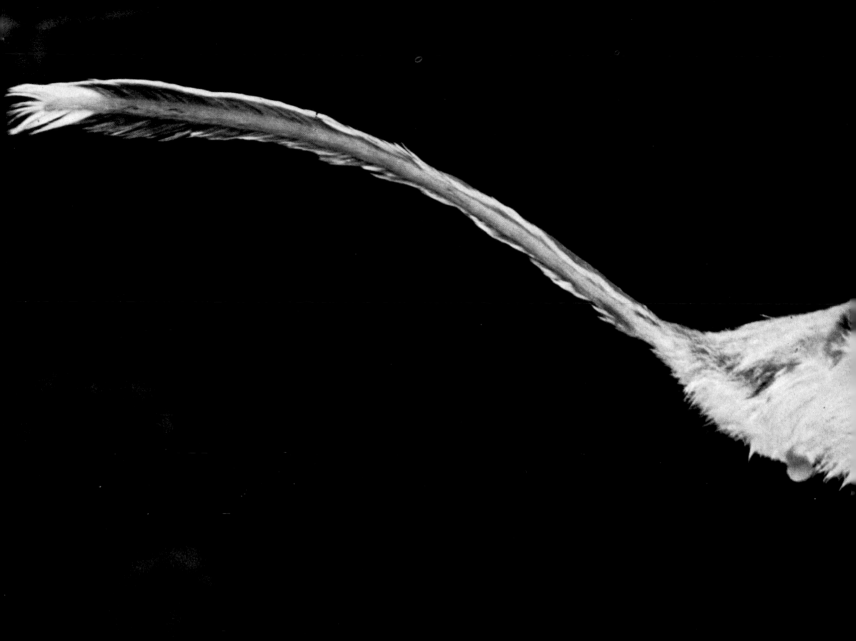

62. *Not a true cat but a marsupial carnivore, the western native cat* (Dasyurus geoffroii) *hunts the Australian forests at night for small mammals and birds, such as the parrot on which it is feeding. The native cat provides an example of what scientists call superfecundation: as many as eighteen young may be born, but only those that find one of the six teats will survive.*

63. *Largest of the surviving marsupial carnivores, the Tasmanian devil* (Sarcophilus harrisii) *is found today only in remote reaches of the island for which it is named. This jet-black, twenty-pound bearlike creature has a massive, muscular head and bone-crushing teeth that enable it to bring down prey as large as a wallaby.*

64 *overleaf. The familiar, slow-motion koala* (Phascolarctos cinereus) *spends much of its time wedged in a fork of a tree—asleep. It has no tail, and it may once have been a ground dweller, but curved claws, long arms, and a viselike grip ease its arboreal life. The koala gives birth to one young, which it raises in a pouch that opens backward.*

66 *second overleaf. At top speed, a red kangaroo* (Macropus rufus) *can dash across the Australian plains at 35 miles an hour, covering twenty-five feet with each leap. If hard-pressed, a female kangaroo's pouch muscles may loosen, and her joey may fall from the pouch and be lost to a predator.*

Photographic credits for the preceding illustrations:

Tom McHugh/Photo Researchers 57
Stanley and Kay Breeden 58
Michael Morcombe 60, 62
Jean-Paul Ferrero 63, 64, 66

Mammals with Wings

Atop a jungled hill called Tamana, in the center of
Trinidad, the earth opens a dark, jagged maw. The forest
growing around the somber breach in the surface is lush
and lively. Pecking after seeds on the ground, tinamous
whistle, icy clear. Cicadas buzz warmly from hidden
arboreal havens. Sunlight filters through the trees and
dapples the red lobster-claw bracts that sheathe the tiny
flowers of the *Heliconia* plants. Within that hole in the
hilltop, however, the colors and music of the forest are
extinguished, the light obscured by a lip of rock just
below the rim. A dozen feet down, dimly seen and alive
with tiny frogs, a slick, muddy floor slopes abruptly
into abysmal gloom.

And from that gloom, should one trespass there, arises
a sound like a whirlwind. It swirls up, as though from the
black bowels of the planet, the sound of bad dreams and
childhood fears. Vast and pervasive, the sound becomes
one with the darkness, and as it mounts to a dry, leathery
thunder, its source becomes clear. Wings make that
sound, countless thousands of them, wings of mem-
branous delicacy, moved by the cave's population
of bats.

The sound of bat wings was first made more than 50
million years ago, at the dawn of the Age of Mammals,
when bats very similar to those living today evolved
the capability of powered flight. No other group of
mammals has ever repeated that feat. For this, if no
other reason, bats ought to be admired, yet they are
abhorred almost universally by humans. A major
exception seems to be the Chinese, who view bats as
beneficent—the Chinese word for bat, *fu*, sounds

identical to the character meaning happiness or
good luck.

Western aversion to the bat stems from manifold roots.
The probable reason that medieval artists equipped
Satan and his demoniac minions with bat wings, for
instance, is that many kinds of bats roost in caves. Some
bat caves, astoundingly, are tenanted by millions of the
winged mammals, hanging upside down by the feet or
wedged into crevices. Their droppings, falling
year after year, may pile up several feet high
on the cave floor.

It did the image of the bat no good when Cortez
encountered species in tropical America that fed upon
blood. The notion that a human bloodsucker could take
the guise of a bat was quickly grafted upon the
European vampire tradition. Because of the unsavory
associations that have developed around it, the bat has
been treated with a loathing directed at no other mammal
and usually reserved for scaly or many-legged things
that creep and crawl.

Not surprisingly, fear has blinded people to the truly
splendid creature that is the bat, especially in view of
the spectacular diversity of its more than 800 different
species. Within this unique assemblage of winged
mammals, there are myriad variations upon
the basic theme, in mode of living, appearance,
and size.

Giants among the bats are the Old World fruit eaters
(Pteropus), some of which carry their rabbit-sized bodies
on wings measuring almost five feet from tip to tip.
From such strapping creatures, bats range in size all
the way down to the minuscule Philippine bamboo bat
(Tylonycteris pachypus), which has a body less than
two inches long. This winged Lilliputian roosts snuggled
within the hollow of a bamboo stem, clinging to the walls
of its refuge with suckers on its feet. It belongs to a
vast family, almost global in distribution, which includes
the North American little brown bats *(Myotis)*, big
brown bats *(Eptesicus fuscus)*, and pipistrelles
(Pipistrellus). They are the bane of insects that fly at
dusk and by night, which the bats pursue, wings a-blur,
in zigzag chases through the air. Insects, however, are
not the only creatures that fall prey to bats; some bats
eat such unlikely victims as shrimp, frogs, and even
one another.

The bulldog bat *(Noctilio leporinus)* of the American
tropics skims the surface of ponds, streams, and even

the sea, with its legs extended and long claws trailing an inch below the surface. Repeatedly, it makes sweeps of several feet, each in the space of a second, until its claws snag a small fish, which is eaten on the wing. The spear-nosed bat *(Phyllostomus hastatus)*, roosting on the roof of the Tamana Hill cavern, issues forth at dusk in large flocks, and seasonally gathers at groves of sapucaia nut trees to gobble their ripened seeds. Most of the time, however, it feeds on flesh, and regularly kills birds, rodents, and smaller bats. So does its relative, the false vampire *(Vampyrum spectrum)*, which swoops out of the nocturnal hush to gaff its victims with the hooked talons that tip its thumbs. The largest bat of the Americas, the false vampire kills quickly, with a skull-crushing bite, and it consumes the body of its victim, not just the blood, as once was thought.

Blood alone is the diet of the true vampires, however, three species that belong to the family Desmodontidae. These bats can digest no other food. They feed, not by sucking, as commonly believed, but by lapping up the blood with feverish speed. When feeding, the vampire inverts the edges of its tongue, which when placed over a groove in the lower lip forms a channel through which the blood flows. The motion of the tongue may even create a slight vacuum, easing resistance to the flow. The wound from which the vampire draws its sustenance is slight, a tiny scoop of skin nicked out by razor-edged upper incisors. So deft and painless is the cut, it seldom awakens sleeping victims, human or animal. In an entire night's feeding, a single vampire consumes only an ounce or two of blood, which is hardly disabling to the host, although farm animals are sometimes weakened by repeated forays of more than one bat. The real danger from the vampire's bite, however, is far more damaging than the loss of blood, for it is one of the bats that often carry rabies.

More palatable, perhaps, to the human way of thinking are the feeding habits of the flower bats. The nectar-feeding bat *(Anoura geoffroii)* flits through the jungles of Mexico and South America, stopping to tap flowers for their sweet fluid. It can get at the nectar of even deep, vaselike blooms, for its tongue is longer than its four-inch body. When the towering saguaro cactus of the Sonora Desert blooms in the darkness, it is visited by another nectar feeder, the long-nosed bat *(Leptonycteris nivalis)*, which in the course of feeding spreads the pollen of the giant plant. Blossoms also draw some of the huge Old World fruit bats, which unfortunately do not merely drink but mash the blooms to a pulp. Major agricultural pests in some parts of the world, the fruit bats chew the juice from blossoms and fruit and spit out the rest.

The giant fruit bats, in addition to size, are characterized by their long-snouted, doglike faces, and eyes that are large and luminous. Internally, they share another trait which, along with their big eyes, reveals an important fact about the way they live. The part of the brain devoted to vision occupies approximately the same amount of space as the portion concerned with hearing —in direct contrast to the other bats, in whose brains auditory areas have expanded vastly at the expense of those concerned with vision. The differences in brain structure relate directly to the fact that, with the few exceptions that always seem required by nature, fruit bats orient themselves visually, whereas other bats use sound. Indeed, most bats have a natural sonar that enables them to navigate by what scientists call echolocation. The bats literally talk themselves toward a target, or around obstacles, by assessing the echoes of high-pitched sounds broadcast by their larynxes and bounced off the target. As they cruise about, bats often broadcast a steady series of staccato, pulsing sounds, but when engaged in aerial pursuit or other tricky maneuvers, they may emit pulses at the buzz-saw tempo of 200 per second.

The wings that carry bats through the air are composed of only two layers of skin, tightly framed upon limbs resembling delicate, elfin arms. Powered by muscles in the chest, the seemingly fragile wings can loft bats as high as 10,000 feet and take them on seasonal migrations of 1000 miles. Some other mammals, as we have seen, can sail through the air, but ultimately they are bound to the earth. The bats, however, by virtue of having true wings, have added another dimension to the mammalian domination of the earth. They have conquered the skies.

73. *Extraordinary among an extraordinary order of mammals is the common vampire bat (Desmodus rotundus), which inhabits both deserts and tropical forests from northern Mexico into Chile and Argentina. Its diet is exclusively blood! On its nocturnal foray, a vampire bat will suck a tablespoon of blood from its chosen victim— a cow, horse, burro, deer, peccary, or on infrequent occasions a sleeping human—then retire to a cave or hollow tree to digest its meal. A vampire alights on the ground and crawls to its victim, cuts away a thin slice of skin with an almost painless bite, then forms a suction tube with its tongue and its deeply grooved lower lip. The bat's saliva is believed to be an anticoagulant that keeps the blood flowing freely for the half-hour feeding period. Neither the bat's bite nor the loss of blood is serious, but vampires can transmit rabies and livestock diseases.*

74 *overleaf. Having roosted too close to each other, two African leaf-nosed bats (Hipposideros commersoni) attack with bared fangs, flailing arms, and ultrasonic shrieks. Damage from such fights, however, is rare. Beetle larvae, dug out of the pulp of wild fig fruits, are a mainstay of the diet of this species. Leaf-nosed bats have flaps on their nostrils that, scientists believe, allow them to emit ultrasonic sounds with their mouths closed. Thus they do not need to cease echolocation when they eat insects snatched in flight.*

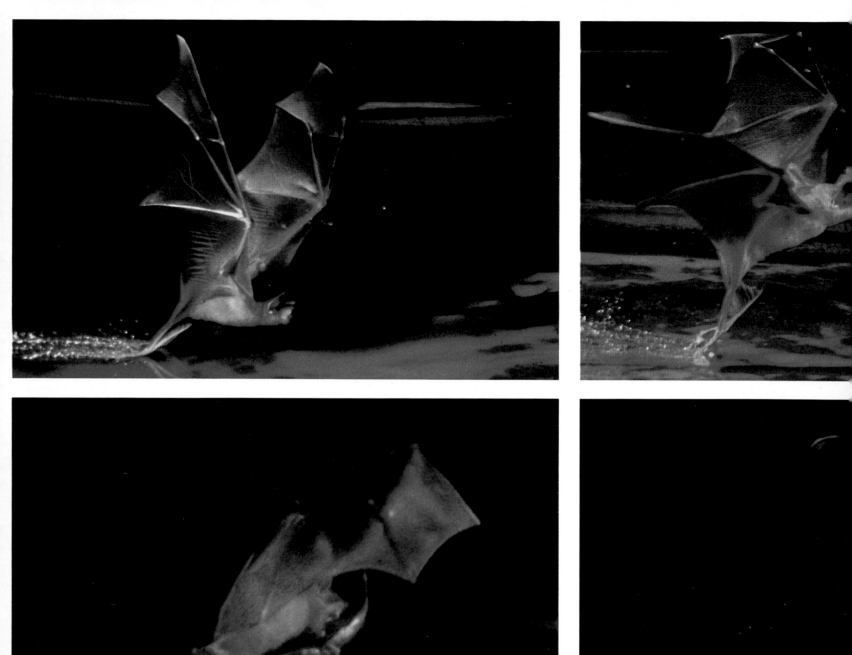

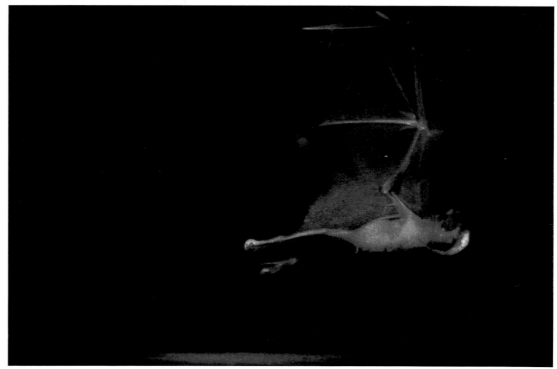

76-77. *The bulldog bat* (Noctilio leporinus) *of Central and South America is a skilled fisher of both fresh and salt water. Raking the surface with long claws, it gaffs a small fish, swiftly lifts the catch to its mouth, and either eats it on the wing or chops the fish into pieces, storing them in its cheeks for later consumption. In late afternoon, numbers of chirping bulldog bats may be seen zigzagging over the surf in the company of brown pelicans, gleaning small fish stirred up by the big birds. Scientists still debate whether bulldog bats use echolocation in tracking fish, but it seems likely that, hunting in areas where fish are plentiful, they simply snare them at random.*

78 *overleaf. At the end of a downstroke, the wings of the Egyptian fruit bat* (Rousettus aegyptiacus), *which span nearly two feet, are fully extended and symmetrical. Bats of the genus* Rousettus *inhabit the Old World tropics from Africa to Southeast Asia and on to the Solomon Islands, feeding on nectar and the juice of fruits, especially wild figs. They are predominantly cave-dwellers, but the Egyptian fruit bat often is found roosting in ancient tombs and temples.*

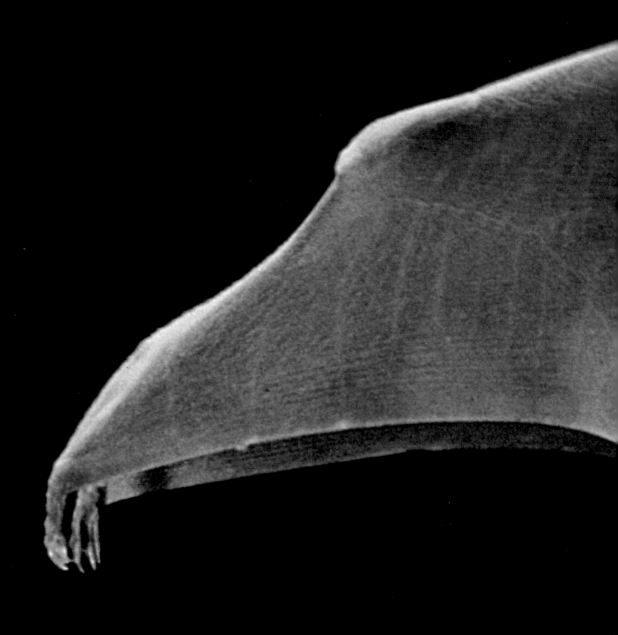

80. *An elongated snout and a long tongue, tipped with a brush of hairs, enable a hovering long-nosed bat* (Leptonycteris nivalis) *to reach deep within the flowers of an agave in search of pollen. Hundreds of thousands of long-nosed bats migrate between Arizona and Mexico, following the blooming season of desert plants.*

81. *By ramming its snout into the corolla, an impatient long-nosed bat has forced open a flower on the tree-like saguaro cactus. Bats are important pollinators of the night-blooming saguaro, which in turn is an important source of nourishment for the bats. A single blossom, jutting safely beyond the spines of the cactus, can produce a tablespoon of sugary nectar.*

82 *overleaf. Crammed together in a Kentucky cave, thousands of little brown bats* (Myotis lucifugus) *hibernate through the cold winter months. Many of them flew hundreds of miles to reach this site. The bats of the genus* Myotis, *often called mouse-eared bats, are found everywhere in the world except Arctic and Antarctic regions and some mid-ocean islands. They hunt exclusively for insects, resting after a feeding flight to digest their catch.*

Photographic credits for the preceding illustrations:

Gnawing Hordes

With typical rodent resourcefulness, a muskrat has built her nursery under the overturned hulk of a rowboat rotting beside a small backyard pond. Her five young hide beneath the old boat, their small bodies huddled together in a mound of glossy fur. Quietly, they await her return from feeding on the leaves of irises which the pond's human owner has planted along its margins. Something approaches their shelter, something that is not their mother, but considerably larger, and heavy of foot. Daylight floods the interior of the nursery as the bow of the boat is lifted from the ground. Instantly, the heap of fur separates into five small forms which scatter in headlong flight. Their dispersal seems frantic, wildly haphazard, but actually the flight of the muskrats is entirely purposeful, the paths deliberate. Although the young rodents scurry in many different directions, each route leads to the same goal, the water. The water is safety, security, and concealment, a refuge which ages of evolution have programmed them to seek when threatened.

The way the muskrat *(Ondatra zibethica)* has so fully exploited a niche in the waters of ponds, lakes, streams, and wetlands demonstrates the remarkable breadth of rodent adaptability. In their astonishing success, rodents have become nearly global in range, and have adapted to virtually all habitats open to mammals except the sea and the air, although, to be sure, some, such as the American flying squirrels *(Glaucomys)*, make brief, unpowered forays into the sky. The rodents display remarkable variety and profusion, as well as tenacity. The world over, they are harried by predators—furred, feathered, and scaled. A host of hungry creatures, from rattlesnakes to lions, feasts upon the rodents, and humans war upon them; nevertheless they persist or, more correctly, flourish. Some, such as the Utah prairie dog *(Cynomys parvidens)* and the Australian thick-tailed rat *(Zyzomys pedunculatus)*, are extremely rare. But many others, as humans have learned to their sorrow, are so abundant that the rodents as a group constitute the majority of mammals living on this planet. In species alone the rodents are legion, their varieties numbering in the thousands.

Within this immense multitude are creatures as disparate as the semi-aquatic capybaras *(Hydrochoerus)*, tropical American beasts that can weigh more than 100 pounds, and the tiny Cape mole rat *(Georychus capensis)* of Africa, which tunnels after tubers and roots. Rodent habits run the gamut of possibilities. The giant naked-tailed rat *(Uromys caudimaculatus)* of New Guinea climbs trees in search of coconuts. The North American pocket gophers (Geomyidae) burrow under plants and pull them down into their underground dining chambers. The pudgy mountain beaver *(Aplodontia rufa)* emerges from its burrows to feed upon ferns and pine needles on the well-watered slopes of mountains in north-western North America. The water rat *(Hydromys chrysogaster)* of Australasia kills and eats fish, crustaceans, lizards, birds, and even other rats.

Despite the seemingly great difference between them, rodents differ only superficially. No matter where and how they live, all have two pairs of curving incisor teeth, chisel sharp, which grow throughout their lives. The incisor of a rodent is a marvel of natural engineering. The part seen in the front of the rodent's mouth is only the tip of the tooth, which actually is rooted far back in the jaw. One pair juts out from the lower jaw, the other directly above, so that in normal use they wear against each other, keeping the edges of the teeth keen. Were it not for the wearing action, the tooth would continue to grow in circular fashion, until it has penetrated the skull or jaw. Because only the front of the incisor is enameled, the wear bevels the tooth from front to back, creating the mark of the rodent group. Gnawing incisors are also the trademark of another group, successful in their own right but not on such an impressive scale as the rodents. These animals, the rabbits and hares (Leporidae) and the little pikas (Ochotonidae), have more than one pair of chisel teeth in the upper jaw. Born with three sets of upper incisors, they retain two pairs as adults, although only one pair does the work, as in the rodents.

With its incisors, the rodent can gnaw through the hardest shells, seed hulls, and rinds to get at the soft food within. Acorns, hickory nuts, and even snail shells present no obstacles to rodents eager to consume their contents. The nemesis of man, the brown rat *(Rattus norvegicus)*, can chew through metal sheeting or cinder block to get at human food stores. Using the upper incisors as a brace against the trunk and shaving away at the wood with the lower pair, a beaver *(Castor)* can fell a tree in minutes.

The sharp-edged incisors also serve as a defensive weapon, which can be wielded with murderous results. The vicious bites of a cornered rat have made more than one attacker regret the choice of opponent. The hamster *(Cricetus cricetus)* is known for its slashing attacks when aroused. Generally, however, rodents prefer not to do battle but to rely for safety on their ability to remain inconspicuous. As is known by anyone who has a pet hamster or gerbil *(Meriones)*, or whose house walls are a winter refuge for scurrying troops of deer mice *(Peromyscus maniculatus)*, it is the night that brings forth the rodent hordes. In southern Africa, white-tailed rats skitter through the darkness over open areas, foraging for seeds. On the deserts of western North America, kangaroo rats *(Dipodomys)* emerge from their labyrinthine burrows to dance like elves upon the sand. In the swamps of South America, fish-eating rats *(Ichthyomys)* course along the waterways. In northern Eurasia, night is the time when mole rats *(Myospalax)* make their infrequent appearances on the surface.

There are a few rodents which, unlike their kin, sleep at night and are abroad by day. The woodchuck *(Marmota monax)* and the hoary marmot *(M. caligata)*— a large ground squirrel—emerge from their cavernous burrows to feed by day and even sun themselves for hours at a time, seldom far from their underground havens. Hoary marmots, gregarious creatures that live in colonies from Alaska to Idaho, rely upon one of their number to serve as a sentinel when they are feeding. The sight of an eagle or a coyote will evoke a shrill warning whistle from the sentry, which sends all the marmots scurrying for their holes. Similarly, the earsplitting crash of the spatulate tail of a frightened beaver slapping the water will trigger a rush away from the shore and shallows by all beavers within hearing. Rats and mice have an extensive repertoire of alarm calls, mostly uttered at frequencies too high to be received by the human ear.

When it comes to the survival of the rodents as a group, however, it is not the ability to keep from being eaten that really matters, but rather their reproductive powers, which are unmatched among mammals. They begin breeding when very young, produce several offspring after a short gestation, and often deliver more than one litter a year. Consider the cotton rat *(Sigmodon)*. In its usual life span of about a year it may produce several litters of up to a dozen young each. Among the equally fecund rabbits and hares, the American cottontails *(Sylvilagus)* produce up to five litters of about a half dozen young each in the space of a year. The showshoe hare *(Lepus americanus)* can deliver almost a score of young per year, although it usually produces only about half that number. The true rabbit *(Oryctolagus cuniculus)* of the Old World, ancestor of all domestic varieties, is fertile for more than a dozen years, if allowed to live that long, and on the average has more than ten young every year.

Not surprisingly, in view of their reproductive abilities, various species among the gnawing hordes experience cyclic population bulges, separated by lows resulting from epidemics or other calamities. Lemmings *(Lemmus)* multiply immensely over a cycle of three or four years, until their food supply no longer can sustain them. The buildup of lemming populations triggers mass migrations of the small rodents, particularly in Scandinavia. Driven by an unyielding urge to move on, a flood of lemmings sweeps out of the hinterlands in a mindless crush that apparently has no goal but to keep traveling. For most of the lemmings in the swarm, the journey ends only in the claws or jaws of predators, or in the chill waters of rivers or lakes or, at long last, the sea. In the end, nature achieves a balance.

89. *Scourge of civilization, the house mouse* (Mus musculus) *has been transported everywhere that man lives, carrying human diseases, destroying crops, ruining food stores, invading homes. Yet this same hated pest, in its albino and domestic forms, is an invaluable tool in medical research, especially in the search for the causes of and a cure for cancer.*

90 *overleaf. The golden mouse* (Ochrotomys nuttalli) *is a white-footed mouse of the southeastern United States that has chosen an arboreal life. It builds a six-inch-wide nest several feet off the ground in brush or the fork of a tree, and the nest may be used by many generations over several years. A second nest-like structure nearby is the mouse family's eating place.*

91 *above left. An inhabitant of coniferous forests and tundr ross northwestern Canada and Alaska, and from Siberia to Norway, the northern red-backed vole* (Clethrionomys rutilus) *is active day or night the year around, gathering and storing tender vegetation, seeds, bark, lichens, fungus, and insects.*

91 *above right. Best known of some fifty-seven species of white-footed mice, the deer mouse* (Peromyscus maniculatus) *is at home in a wide variety of ecosystems across most of North America. Nocturnal in its habits, it has a tail as long as its body.*

91 *center. Large ears, long narrow hind feet, and a tai   o eleven inches long characterize the kangaroo mice* (Notomys sp.) *that inhabit the grasslands, plains, and dunes of Australia. Kangaroo mice share the same habitat—and occasionally the same tunnel systems— with mouselike marsupials.*

91 *below left. The pika* (Ochotona princeps) *of the Rocky Mountains bustles about all day, putting away stores of "hay" for use when winter snows bury the high meadows. It lays out fresh-cut grass for the sun to dry, then tucks it away in rock niches. This is one of two North American representatives of a genus that has fourteen species scattered across the mountains of Asia.*

91 *below right. The world's most primitive rodent is the mountain beaver* (Aplodontia rufa) *of the rain- and fog-soaked coastal forests from northern California into British Columbia. Sometimes known by the name given it by Chinook Indians, sewellel, the mountain beaver is not closely related to a beaver, does not behave like a beaver, and rarely lives high in the mountains. About fifteen inches long, it resembles a tailless muskrat, digs elaborate burrow systems, and eats about any kind of plant material, from berries to ferns to evergreen needles.*

92 *overleaf. Fighting over a small piece of desert, African ground squirrels* (Xerus erythropus) *lash out with their hind feet. Because bits of dirt cling to their hairs, these burrowing animals—widely distributed south of the Sahara—take on the color of the soil where they live.*

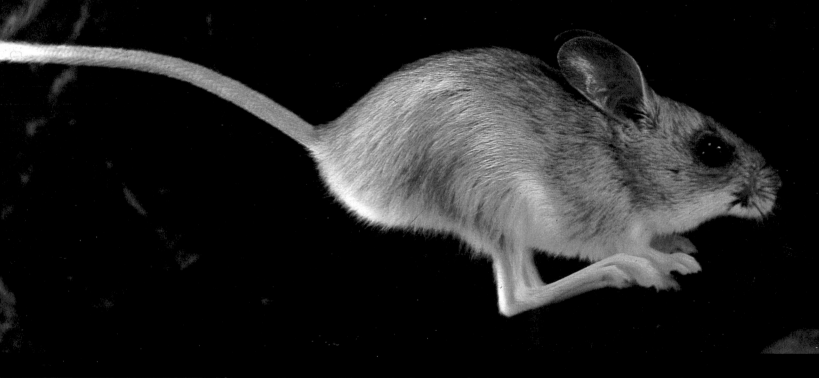

94. *Biggest squirrel in North America, the fox squirrel* (Sciurus niger) *may weigh almost three pounds in autumn, grown fat on the ripe mast crop of acorns, hickory nuts, beechnuts, and walnuts. To tide it over the lean winter months, it buries hundreds of nuts, one at a time, in small holes beneath the trees from which they fell. The fox squirrel varies greatly in color, from bright orange to jet black.*

95 *top. The lodge of the North American beaver* (Castor canadensis) *may be thirteen feet in diameter and reach two feet above the level of the pond. Entered from underwater, the lodge contains a dry sleeping platform and is shared by the parents, the offspring from the previous year, and the latest litter, usually numbering three.*

95 *above. Though aquatic in nature, with partially webbed hind feet and a laterally flattened, nearly naked tail that is used as a rudder, the muskrat* (Ondatra zibethica) *is a close relative of the terrestrial voles. Equally at home in fresh or salt water, in marshes, lakes, or rivers across most of North America, the muskrat can remain submerged for twelve minutes.*

with stiff hair. Strong nails dig through frozen snow to expose twigs and roots of willows, sedges, and saxifrages, which are extracted with projecting incisors that operate like forceps. In storms, the Arctic hare digs a tunnel in the snow for shelter. On Canada's high Arctic islands, these gregarious animals form herds numbering several thousand.

Photographic credits for the preceding illustrations:

The Weasel Clan

In the white silence of a winter woodland, a sinuous
pursuit ends lethally, as with blurring speed a long-
tailed weasel rockets into a fleeing cottontail and kills
it with a quick thrust of needle-sharp teeth to the brain.
High in a tree overlooking a wooded savanna in Africa,
a honey badger, flat and brutish of face, rakes open a
beehive with its hooked foreclaws and gorges on the
soft pupae and the sweet contents of the combs. On the
windswept high plains of the Andes, a hog-nosed skunk
shuffles among the sparse vegetation, naked snout
searching the ground for insects to eat. The weasels
and their relatives, collectively known as the mustelid
family, are among the most diverse of all the carnivores
in the myriad ways they exploit their surroundings to
make a living. The forests, deserts, waters, and plains
of all the continents except Australia and Antarctica
are inhabited by one or another of the seventy species
in this group, a family characterized by a low-slung
body build and by scent glands that produce a rank,
even vile, odor.

The members of the weasel clan also share several other
traits, notably astonishing strength for their size,
tenacious ferocity, and lethal agility. It is these qualities
that are responsible for the group's multi-faceted success.
Their fearlessness, murderous persistence, and power
are legendary. The wolverine *(Gulo gulo)*, at fifty pounds,
is perhaps the strongest of all mammals for its size. Fond
of carrion, it chases cougars and even bears from their
kills. When it needs fresh meat, the wolverine—which
since the ice ages has retreated to the northern high
country and the Arctic fringes—sometimes kills moose

and elk mired in snow. Deer caught in heavy snow occasionally are the prey of the North American fisher *(Martes pennanti)*, a dozen pounds of speed and ferocity, which is also the bane of porcupines. Once it corners a porcupine, a fisher performs like a lightweight boxer. It dances and it feints, dodging the crippling blows of the porcupine's quilled tail, and darting in to repeatedly slash the porcupine's unprotected face. When the porcupine is helpless and dying, the fisher overturns it and tears open its soft underparts.

The African striped weasel *(Poecilogale albinucha)* fences in a similar fashion with venomous snakes, waiting until the reptiles are exhausted before biting into the back of their heads. The badger *(Meles meles)* of Eurasia will squirm around inside its baggy but tough skin and bite back viciously if a dog grabs it by the scruff of the neck. With a grating snarl, it explodes from its burrow to assault trespassers upon its territory, especially in the breeding season. Animals as large as the African buffalo have been fatally wounded by the honey badger *(Mellivora capensis)*, which shreds their groin and genitals from below. Such courage and savagery are typical of the weasel clan, and even the smallest members, such as the least weasel *(Mustela ruxosa)*, which weigh only a few ounces, have held their ground when men have approached them on a kill.

The manner in which weasels *(Mustela)* make their kill is a chilling exercise in relentlessness. Like a slim missile homing in on a target, a long-tailed weasel *(M. frenata)* will match every evasive twist and turn of a rodent or hare, as though linked to its prey by an invisible strand. Almost invariably, the chase ends with a fatal strike at the base of the victim's skull, or the neck, often delivered with blinding speed. The lithe, slender body of the weasel enables it to prowl the underground burrows of rodents with almost serpentine ease. Virtually anywhere a weasel can wedge its pointed black snout, it can take its body. Mice are tracked to the ends of their burrows, and knotholes serve as entrances to henhouses. Once inside, the weasel may slaughter as many hens as it can reach, but not because it is particularly bloodthirsty. When a weasel is confined with an abundance of prey, its highly tuned predatory responses may be difficult to shut down, and it may be unable to stop killing as long as the prey is moving.

Even trees offer no escape for the fleeing victim of a weasel, for it will go aloft after its dinner. Many members of the clan, in fact, display the same lightness of foot in the branches as on the ground. The martens *(Martes)* do most of their hunting in the trees and subsist largely on squirrels. The tayra *(Tayra barbara)*, a yard-long beast of the American tropics, takes to the trees to hunt arboreal anteaters and steal bananas. If pursued by men with dogs, as is likely after a raid on a banana plantation, the tayra may use the forest canopy as a highway. Even the ground-dwelling ferret badger *(Melogale moschata)* of Formosa sometimes forsakes its burrow to sleep cradled in the boughs.

Of all environments, however, the one in which the mustelids truly surpass the other carnivores is in the water. The river otters *(Lutra)* and the minks—*Mustela vision* of North America and *M. lutreola* of Eurasia— manage the weasel-like dazzling maneuvers in the water, as they swim with supple grace and swiftness after fish. They have established a virtual monopoly over the role of freshwater mammalian predator. And two of the mustelids even have followed evolutionary paths which have led to the sea, specifically the Pacific Ocean. The small marine otter *(Lutra felina)* catches fish and other seafood where the chill waters of the Humboldt Current sweep close to the west coast of South America, from northern Peru south. However, the marine otter still retains substantial links with land, for it eats its catch ashore and dens there. The endangered giant otter *(Pteronura braziliensis)* of South America is the longest—six feet from snout to tip of tail. Weighing hardly more than an ounce, the least weasel is the bantam among the group and, in fact, is the smallest carnivore.

Big and small, mustelids have highly developed scent glands whose secretions range in potency from slightly pungent to overpowering. The skunks, of course, are notorious for the chemical warfare they wage on their enemies. Some of their mustelid relatives, however, are just as odorous. The green secretion shot by the aptly named Malayan stink badger *(Mydaus javanensis)* is supposedly potent enough to asphyxiate dogs which receive a substantial dose. The odor given off by the marbled polecat *(Vormela peregusna)* of Eurasia and the African zorille *(Ictonyx striatus)* is also devastating. Most of the mustelids wear some sort of contrasting blotches or stripes which warn other beasts to give them a wide berth. As a rule, the more powerful their chemical defense, the more vivid their markings. The impact of such weapons has not been lost on humans. Witness the epithets inspired by such creatures as skunks and polecats. As a group, in fact, the mustelids probably have been associated with more unsavory human actions than any other mammals. At the same time, they have been among the most prized of creatures, valued so greatly for their furs that men have risked their lives in the wilderness to trap them. Ermine, the winter white-coated weasel, was once reserved for royalty. Marten, commercially known as sable, and mink are still reserved largely for the wealthy. What is most priceless about the members of the weasel clan, however, is their magnificent savagery, symbolic of a wild world that is rapidly becoming only a memory.

105. *At home in every kind of land habitat from southern Canada into South America, the long-tailed weasel* (Mustela frenata) *is a super predator in a small and attractive package. No more than sixteen inches long, it is normally chocolate brown with a yellowish belly, but in northern areas it turns pure white in winter—except for a black tip on its six-inch tail. Furriers call winter weasel pelts "ermine," and in medieval Europe they were reserved for the garments of royalty; 50,000 skins of the short-tailed weasel or stoat were used for the coronation of King George VI of Great Britain in 1937. Such variable beauty disguises a fierce hunter of squirrels, rabbits, snakes, frogs, insects, songbirds, and a multitude of mice and rats. Indeed, the weasel has been accused, not without cause, of killing just for fun; a weasel that finds itself in a henhouse, for instance, may slaughter dozens of chickens in a single night. One scientist called it "the most bloodthirsty of all animals," adding that "its favorite drink is warm blood sucked from the neck of its prey." And a weasel will not hesitate to attack a man who gets in its way. But this petite killer is not without its own mortal enemies: hawks, owls, and even prowling house cats.*

106-107. *In a remarkable sequence of photographs, a long-tailed weasel sights a cottontail rabbit in the snowy mountain forests of Wyoming and launches an attack with blurring speed almost too fast for the human eye to follow. Although the rabbit is many times larger than the weasel, its size gives it no advantage over its bantam adversary, and the outcome of the pursuit is inevitable. Like a slim white missile homing on a target, the weasel stays with its chosen victim as it frantically twists and turns in the snow, finally seizing the cottontail at the base of the skull and sinking needle-sharp teeth into the brain. The hunt ended, the seven-ounce weasel drags the three-pound rabbit off into the forest.*

109. *Although they are lovely to look at, the skunks of the New World are not animals to be disturbed. An arched white-plumed tail is a fair warning to intruders that the skunk is prepared to spray—with remarkable accuracy—a foul-smelling fluid that burns the eyes and whose acrid odor persists for days. Most frequently encountered is the striped skunk* (Mephitis mephitis), *which has the distressing habit of taking up residence beneath suburban homes and summer cabins.*
110 *overleaf. Four feet long, weighing perhaps fifty pounds, and suggesting a small bear, the wolverine* (Gulo gulo) *fears no other creature. It will kill an elk or moose struggling in deep snow, drive a mountain lion or grizzly bear away from its meal, attack a bear cub, and rob man's traplines. The wolverine roams the tundra and subarctic forests around the pole; a male wolverine requires a territory of about 1000 square miles, which he will share with two or three females.*

108 *top. An American badger* (Taxidea taxus) *devours a rattlesnake. Using its powerful front legs, which are armed with long claws, as well as its strong jaws, a badger can dig a hole in the ground, vanish, and plug the burrow behind it in a few seconds. The badger's coarse fur once was used to make shaving brushes.*

108 *above left. An inhabitant of streams, lakes, and tidal marshes across most of North America, the mink* (Mustela vison) *is a deadly hunter of muskrats. A mink will corner one of these big aquatic rodents in its bank burrow and quickly dispatch it; or rip open a muskrat's cattail lodge, devour the young, and then claim the house as its own den.*
108 *above right. A large weasel that haunts the treetops and is equipped with strong claws for climbing and a long bushy tail to keep its balance, the American marten* (Martes americana) *pursues its favorite prey, the red squirrel, through conifer forests from Newfoundland to Alaska.*

Ancient and Unusual

When the platypus was first brought to the attention of European scientists almost two centuries ago, in the form of a skin, they considered it a fraud. Their error can be excused, for the platypus is so bizarre that even today it is difficult to believe such a creature exists and is, indeed, rather common in some of the lakes, ponds, and streams of eastern Australia and Tasmania. The platypus has a look of ultimate incongruity. Its broadened snout superficially resembles the bill of a duck. Its tail is flattened like a beaver's, and its webbed feet approximate those of an otter. The hind feet of the male, moreover, carry spurs which, totally out of character for a mammal, are venomous. The female, in an even more astonishing departure from the mammalian norm, lays eggs, reptilian in aspect. The platypus, in fact, typifies a time when very possibly all mammals were oviparous and almost as reptilian as mammalian in nature. Perhaps the strangest of mammals, the platypus *(Ornithorhynchus anatinus)* is nonetheless rivaled in oddity by many of the other creatures grouped here.

The oddities include the tree sloths (Bradypodidae) of tropical America, which live a slow-motion existence hanging upside down and sometimes appear green in color because of the algae growing on their hair. The sloths' relatives, the armadillos (Dasypodidae), and other armored creatures also belong to the group. So do the echidnas (Tachyglossidae)—the only other living egg-laying mammals besides the platypus—and the frantic, furious shrews (Soricidae).

Actually, these and the others grouped here are only strange from a human perspective. In the context of nature, they may be unusual, but are in no sense abnormal. They are not freaks, for as far as nature is concerned, their queer appendages and odd behavior are merely routine manifestations of the many ways in which animals have adapted to varying modes of life. The odd combination of organs in the platypus, for instance, superbly equips it for feeding along the muddy bottoms of the streams and lakes whose banks hold its burrows. The skin of the bill is soft and wonderfully sensitive, permitting the platypus to hunt for worms, insect larvae, and other small aquatic animals by touch, with its ears and eyes closed. The fleshy tentacles that tip the snout of the American star-nosed mole *(Condylura cristata)* serve a similar purpose when the little creature forages along the bottom of streams and ponds, in a dramatic demonstration of how moles can swim as well as burrow. A rat-sized Eurasian species, the Russian desman *(Desmana moschata)*, is almost entirely aquatic, and its elongated snout functions not only as a probe but as a snorkel when the mole swims just below the surface. Underground, even more than underwater, the moles also rely on their tactile, naked snouts to feel for food. Their remarkable sense of touch compensates for their extremely poor eyesight and even, in the case of species such as the eastern American mole *(Scalopus aquaticus)*, the lack of external eyes.

Parallel adaptations for coping with the same sort of life situations link many of the otherwise unrelated animals in the assemblage described here. When fencing with a venomous snake, for instance, both the Indian mongoose *(Herpestes edwardsi)* and the European hedgehog *(Erinaceus europaeus)* ruff up their hairs— modified into spines in the hedgehog—so their opponent may misjudge the length of a strike needed to reach flesh. A miss by the snake leaves an opening for the mongoose or hedgehog to dart in and sink its teeth into the reptile's spinal column, killing it. Dietary reliance upon tropical ants and termites, which must be dug from nests as hard as concrete, has shaped several other creatures from separate evolutionary lines in an equally unusual manner. The anteaters (Myrmecophagidae) of the American tropics, the pangolins *(Manis)* of Africa and southern Asia, the aardvark *(Orycteropus afer)* of Africa, and even the little echidnas of Australia and New Guinea all have large, powerful claws for digging, few or even no teeth, tubular snouts, and tongues of snakelike proportions, sticky with modified saliva, for plucking up ants and termites. The giant anteater *(Myrmecophaga tridactyla)*, a beast of some fifty pounds, has a tongue twenty inches long. The giant pangolin *(Manis gigantea)* of Africa, a sixty-pound animal, has a ten-inch tongue rooted all the way back in its pelvis. The aardvark, which can reach 150 pounds, has a foot-long tongue, and the echidnas, the largest of which is twenty pounds, can stretch their tongues a half foot into an ant nest. The evolutionary paths taken by the pangolins and the anteaters in their respective hemispheres converge even more dramatically in some of their smaller representatives which, unlike the larger types, have taken to the trees. The tamandua *(Tamandua tetradactyla)* and the squirrel-sized silky anteater *(Cyclopes didactylus)* and two species of African tree pangolins have prehensile tails, which provide an extra holdfast while they attack the nests of arboreal ants and termites.

Digging also furnishes some of the creatures that live on termites and ants a means of escaping from danger. The echidnas can vanish in a twinkling, as though swallowed by the earth. Once tunneled into the ground, an echidna can brace itself so firmly in its hole that the strongest man cannot dislodge its squat form. The echidna wedges itself in not only with its powerful feet but with its coat of spines. If for some reason it cannot burrow, it rolls itself into a ball, in effect a living pincushion. Most of the armored creatures that can be classed as unusual opt for one or both of the aforementioned defenses if in trouble. The pangolin, which is covered with scales of modified hair, as if in imitation of a pine cone, locks itself into a ball impossible to unravel. The hairy armadillo *(Chaetophractus villosus)* and the pichi armadillo *(Zaedyus pichyi)* anchor themselves in their burrows with the edges of their body armor, which really is skin-covered bony plates. If caught in the open the hairy armadillo will hunker down under its armor, which touches the ground at its lower edges. The huge giant armadillo *(Priodontes giganteus)*, heavier than a German shepherd dog, curls up in its massive shell, but not as completely as the three-banded armadillo *(Tolypeutes)*. The latter, when frightened, tucks in its head, legs, and tail and rolls into an enclosed sphere resembling a pumpkin.

Some of the porcupines, which share with the echidnas and hedgehogs hairs that have taken the form of spines, use their quills aggressively when attacked by flesh eaters. The Old World porcupines (Hystricidae), which can reach a size of sixty pounds, sometimes bristle their quills and dash into an attacker. The North American porcupine *(Erethizon dorsatum)* swings its quilled tail like a spiked club. The quills of porcupines, needle sharp, detach when their points penetrate an enemy, working their way deep into the flesh, sometimes with fatal results. The quills of the crested porcupines *(Hystrix)* of southeastern Europe and Africa can exceed twenty inches, and are an extreme in spiny armor. Extremes always are bizarre, at either end of the scale, which is one reason why the shrews, some of the smallest of living mammals, must be counted among the unusual. The minuscule musk shrew *(Suncus etruscus)* of Italy, tiniest mammal in the world, weighs less than a dime, and the American pygmy shrew *(Microsorex hoyi)* is hardly larger. As is the case with some other very small mammals, shrews must consume tremendous amounts of food to fill their energy needs, and must refuel with worms, insects, and even other small mammals every few hours. Some of the shrews rely not only upon the legendary ferocity of their attack to down relatively large prey, but also upon a weapon otherwise virtually unknown in mammals—a venomous bite. The American short-tailed shrew *(Blarina brevicauda)* has enough venom in its salivary glands to kill 200 mice. The water shrew *(Neomys fodiens)* of Europe introduces enough venom with its bite to weaken frogs and small fish sufficiently for it to dispatch them. The shrews, moles, hedgehogs, and the other insectivores—notably the tenrecs (Tenrecidae) of Madagascar —are among the more primitive of the placental mammals. The first placental mammals, appearing long before the Age of Reptiles peaked, may have resembled tenrecs, and perhaps shrews. But the links the insectivores have with the past pale before the credentials of the monotremes, whose living representatives are the platypus and the echidnas. Although mammals, they retain so many reptilian features that it is certain they arose long before either the marsupials or the higher mammals and may represent a separate crossing of the mammal-reptile line. This is evidenced not only by the fact that, of all living mammals, only they lay eggs, but also by internal features, such as bone structure, and the reptilian sprawl of their legs. They are living reminders of our scaly, cold-blooded past.

117. *One of a kind in the animal world, the duck-billed platypus (Ornithorhynchus anatinus) of Australia and Tasmania is more closely related to reptiles than to any living mammals. For the female platypus lays from one to three sparrow-sized eggs in a nest of wet leaves deep within a sealed burrow, curling her body about them during the incubation period of seven to ten days. The young platypuses are blind and naked when they hatch, and they will not emerge from the burrow for four months. Twenty inches long and weighing two to four pounds, a platypus is uniquely adapted for its wholly aquatic life. Its body is streamlined, its feet webbed, its flat tail a perfect rudder. And its pliable bill is so sensitive that the platypus, while swimming with its eyes tightly closed, can find crayfish, shrimp, snails, worms, and the larvae of aquatic insects in the bottom mud by touch alone. The platypus also has internal cheek pouches in which it temporarily stores the food it has captured during its one- or two-minute forays underwater. In courtship, the male and female platypuses swim in circles, the male holding on to his mate's tail. Australian colonists decimated the platypus for its fur, but strict protection has brought this remarkable creature back from the brink of extinction. Occasionally, the platypus takes revenge upon man for its former travails. The male's hind limbs are armed with venomous spurs, and the poison can cause weeks of suffering to anyone who mishandles this unique animal.*

118. *The name "pangolin" has its origin in the Malayan word* guling, *which means a long round cushion. When these timid, nocturnal ant- and termite-eaters of tropical Africa and Asia are threatened, they roll into a tight ball and erect their sharp-edged scales. Some pangolins are terrestrial, and can walk about on their hind legs using their long tails for balance. Others, like this African tree pangolin with its baby* (Manis tricuspis), *are arboreal and their tails are prehensile. The armored tails also can be wielded as a wicked mace.*

119. *The entire life of a three-toed sloth* (Bradypus sp.)—*as long as twelve years—may be centered around a single tree in the South American rain forest. To feed, the sloth inches along the branches upside down, slowly pulling young leaves, twigs, and buds into its mouth with hooklike claws. Inexplicably, the sloth always descends to the ground to defecate. The hairs of the sloth's fur are adapted for growing algae, giving the animal a greenish tint that serves as a camouflage during the rainy season.*

120 *top and above. The North American porcupine* (Erethizon dorsatum), *which may reach a weight of forty pounds, has two easily recognized forms. The spines of the porcupine of western forests, from Alaska to New Mexico, are tipped with yellow and surrounded by long fuzzy hairs; those of its eastern counterpart are tipped with black.*

120-121 *top. Few predators are able to attack a Eurasian hedgehog* (Erinaceus europaeus) *successfully, for it buries its vulnerable head, belly, and legs in an impenetrable ball of spines. The spines also cushion the impact when the hedgehog falls or drops from a tree.*

120-121 *above. The prehensile-tailed porcupines* (Coendou sp.) *of Central and South American tropical forests vary in color from almost white to almost black. They are nocturnal and arboreal, and their tails lack spines and are naked on top for curling under rather than over tree branches.*

121 *top. Largest of its kind is the South African porcupine* (Hystrix africae-australis). *But the angry rattling of its eight-inch tail, covered with long white spines, does not deter local hunters, who prize its flesh.*

121 *above. Like the platypus, the echidnas or spiny anteaters of New Guinea and Australia are egg-laying mammals, the female incubating her single leathery-shelled egg in a pocket that develops on her belly during the breeding season. The short-nosed spiny anteater* (Tachyglossus aculeatus) *has only one enemy—aboriginal man, who likes to eat its ant-scented meat.*

122 *overleaf. Until she becomes pregnant again, the female giant anteater* (Myrmecophaga tridactyla) *carries her single young on her back as she shuffles about the swamps and savannas of Central and South America, ripping apart ant and termite nests with her powerful foreclaws and capturing the exposed insects with a long and sticky tongue.*

Photographic credits for the preceding illustrations:

The Hunters: Dogs

Creeping low to the ground, its tail outstretched, its belly almost brushing the earth, the red fox stalks a cottontail. Slowly, sharp nose pointed in the rabbit's direction, it inches imperceptibly toward its prey. A dozen feet away it halts, still concealed by the undergrowth. Its body tenses, muscles bunch, and it explodes out of hiding, darting for its prey, which spurts away, with the fox in swift, silent pursuit.

On the prairie, speckled with flowers, a jackrabbit is running for its life. Behind it courses a coyote, jaws parted, teeth gleaming. Ahead, unknown to the rabbit, another coyote lies in wait, preparing for the kill.

Foam flying from its mouth and nostrils, a sambar deer flees across the landscape of southern Asia. Its flanks are bleeding, its hindquarters ripped open. With it runs terror, silent and tenacious, the dhole pack—wild dogs that are the scourge of the eaters of grass, leaves, and buds. The sambar still lives, but it will die as it runs.

The dog family (Canidae) consists of hunters generally built for the chase, although some members are sufficiently sly to catch prey by stealth. A few, generally only the smaller members of the group, are solitary hunters. The agile, intelligent red fox *(Vulpes vulpes)* prowls the land alone, as silently as a blown leaf, ready to run down a rabbit if necessary, but also eager to pounce upon a mouse that has stirred in the grass. In its hunting methods, it sometimes behaves more like a cat than a dog. The gray fox of North America *(Urocyon cinereoargenteus)* is also a lone hunter. Nightly, it follows the same route, a twisted, winding course over the countryside as it casts about for small rodents. Similarly, the coyote *(Canis latrans)* is prone to patrolling its territory over a regular hunting trail, night after night, even year after year. Yet for the coyote the hunt can also be a family affair, which is why coyotes, seldom exceeding thirty pounds, can sometimes kill prey that is bigger, stronger, and faster than they, and it is this aspect of their behavior that so strongly characterizes the group. The members of a coyote family will take turns running a pronghorn antelope in a circle until it falls from exhaustion, whereupon all the coyotes move in for the kill. The same tactic is used by the small golden jackal *(Canis aureus)* of Africa and Asia to occasionally catch deer and similar hoofed creatures.

At their most devastating the canine hunters work as highly organized, lethally efficient teams, ranging from two to a dozen or more members. Deep-chested, lean, and long of leg, the canine hunters are typified by their relentless pursuit, seemingly ruthless but actually entailing only the degree of violence necessary to make the kill. The evolution of team tactics among them has been accompanied by the growth of social organizations that are so sophisticated and complex they rival those of the most advanced primates. The technique of hunting as a group is most highly developed, and most harrowing to watch, among the wolves *(Canis lupus)*, African hunting dogs *(Lycaon pictus)*, and the dholes *(Cuon alpinus)* of Asia.

These creatures form hunting packs that operate with superb coordination, and regularly kill animals which easily could repel the attack of a single one of their number. African hunting dogs have been known to employ some members of their pack as decoys to distract the victims from the real assault. Wolf packs split up to separate a cow moose from her defenseless calf. A few members of the pack will drive the cow, harrying her so that gradually they split her from the calf, pressing her farther and farther from her offspring, while the remainder of the pack swarm over the helpless young.

The dholes, the feared "red dogs" of Kipling, sometimes engage in elaborate maneuvers, likened to beating the bush for game and running the prey that is flushed into a trap set by part of the pack.

Dholes hunt primarily by scent, an advantage in the thick cover they generally inhabit. Wolves and African hunting dogs, which often live in the open, hunt largely by eye, but if they are seeking prey in the forest or thick bush, they also nose about in an effort to pinpoint the location of their victim. Once it has been targeted, they sneak as close to it as possible before revealing themselves. Then there is a furious rush toward the prey, as the hunters try to bring it to bay before it can flee. Large targets, such as a large antelope, may stand their ground and fight, and more often than is generally believed, drive off their tormentors. But once the victim chooses to flee, the advantage is with the pack, which strings out in pursuit, sometimes running in relays, until exhaustion forces the hunted creature to stop. If the hunters are dholes, the quarry often is half dead before it stops running, for they harry their victims in a particularly gruesome manner. Running alongside, even under the quarry, the red dogs snap and tear at it with crippling bites, literally eating their victim alive as it flees.

Once the quarry of the pack has been brought to bay, the tactics used by all of the wild dogs are the same. If the prey is small, it is immediately overwhelmed in a flood of snarling bodies. Large victims are encircled by dancing, darting forms, trying from every direction for a telling bite. Wolves and African hunting dogs customarily try for a grip on the snout of the victim. When one of the pack has managed to hold fast, the rest attack the quarry from the rear, trying to get it to the ground.

As might be surmised, to cooperate so smoothly, packs must be highly structured and joined by strong communal ties. The pack is, then, a form of extended-family unit, which serves not only to provide food but for protection as well. Youngsters growing up within the pack benefit almost immediately from the ties that hold it together. Wolves and hunting dogs swallow chunks of flesh at the kill, then return to the young and regurgitate it for them to eat. Dholes sometimes form community nurseries, and it is thought that the females may share with each other some of the responsibilities of rearing the offspring.

The size of a pack varies from a few to more than two dozen members, but to work most effectively, it needs at least six animals. Groups of fewer than that are at a tremendous disadvantage, not only during the hunt, but also because they probably are not large enough for the development of the complex network of relationships that cement the pack together as a unit. The maximum size of the pack is rather flexible, for it is most likely linked to external conditions. Where food is abundant, and large prey abounds, the size of the pack can increase without some of its members going hungry after a kill. But if a pack is so large that even successful hunts leave many members still famished, it will break down, perhaps into two new hunting bands, which together can cover a much greater expanse of territory and thus multiply the food supply available to the same number of animals.

Thus, the hunting pack, as formed by wild canines, is a means of exploiting food that would seldom if ever be available, without the strength that comes from unity.

129. *No wild dog is found over such a wide variety of habitats as the dhole (Cuon alpinus) of India and Asia. Although primarily a forest dweller, it is equally at home along seashores and high on earth's greatest mountains, the Himalayas. It will feed on sea turtles, ripping off their shells, and drive mountain sheep over precipices. The "red dog," as it is often called, hunts in packs that may number thirty animals, but because it lacks the great speed of other canids it will doggedly follow a chosen victim—ibex, reindeer, wild boar, sambar— for hours before finally making the kill by grabbing the prey from behind or disemboweling it on the run. Also unlike many other wild dogs that hunt in packs, the dhole is relatively quiet, neither barking nor howling. Biologists report that the dhole hunts across vast territories and rarely makes more than one kill at a time in a particular area. They also note its fondness for the blossoms of wild rhubarb. Once widely distributed from Siberia south across China into Sumatra, Java, and India, the dhole is rare or extinct today over much of its historic range.*

130-131. *African wild dogs (Lycaon pictus) are famous among animal behaviorists for their highly organized social life. Packs of anywhere from a half-dozen to sixty wild dogs live as a close-knit community, and while there is no dominance hierarchy, responsibilities are strictly divided. While some dogs lead the hunt, pursuing wildebeests, warthogs, gazelles, and zebras across the savanna at speeds reaching 30 miles an hour, other members of the pack stand guard at the dens — usually old aardvark holes — where litters of six to eight pups are hidden. When the hunters return, they regurgitate partly digested meat to feed nursing females, young, and aged members of the pack.*

132 *overleaf. A pack of wolves (Canis lupus) pursues a moose — unsuccessfully this time — in the deep snows of Isle Royale National Park, in Michigan. The wolves of this forty-mile-long wilderness preserve in Lake Superior are equally famous in research circles. Breeding wolves first reached Isle Royale in the late 1940s, crossing eighteen miles of ice from the Ontario shore. Their unexpected arrival was nature's answer to a dilemma — a population of moose that, in the absence of any predator, had exploded and severely overbrowsed the forest. Today, two dozen wolves live on Isle Royale in careful balance with a moose herd that in midwinter averages a thousand animals.*

134. *Largest of all wild dogs, a wolf may stand thirty inches at the shoulders and weigh 150 pounds and is capable of bringing down prey as big as musk-oxen and caribou.*
135. *Falling snow dusts the thick pelt of a red fox* (Vulpes vulpes) *sleeping off a heavy meal. On Isle Royale in winter, scavenging foxes are a principal beneficiary of moose kills by the wolf pack.*

136 *overleaf. The gray fox* (Urocyon cinereoargenteus) *of North and South America has remarkable habits for a member of the dog family: it climbs trees and likes to eat ripe fruit and grain.*
138 *second overleaf. Speedster of the North American prairies, the coyote* (Canis latrans) *can run 40 miles an hour. Although it is blamed by ranchers for killing sheep, its primary diet consists of rabbits and rodents.*

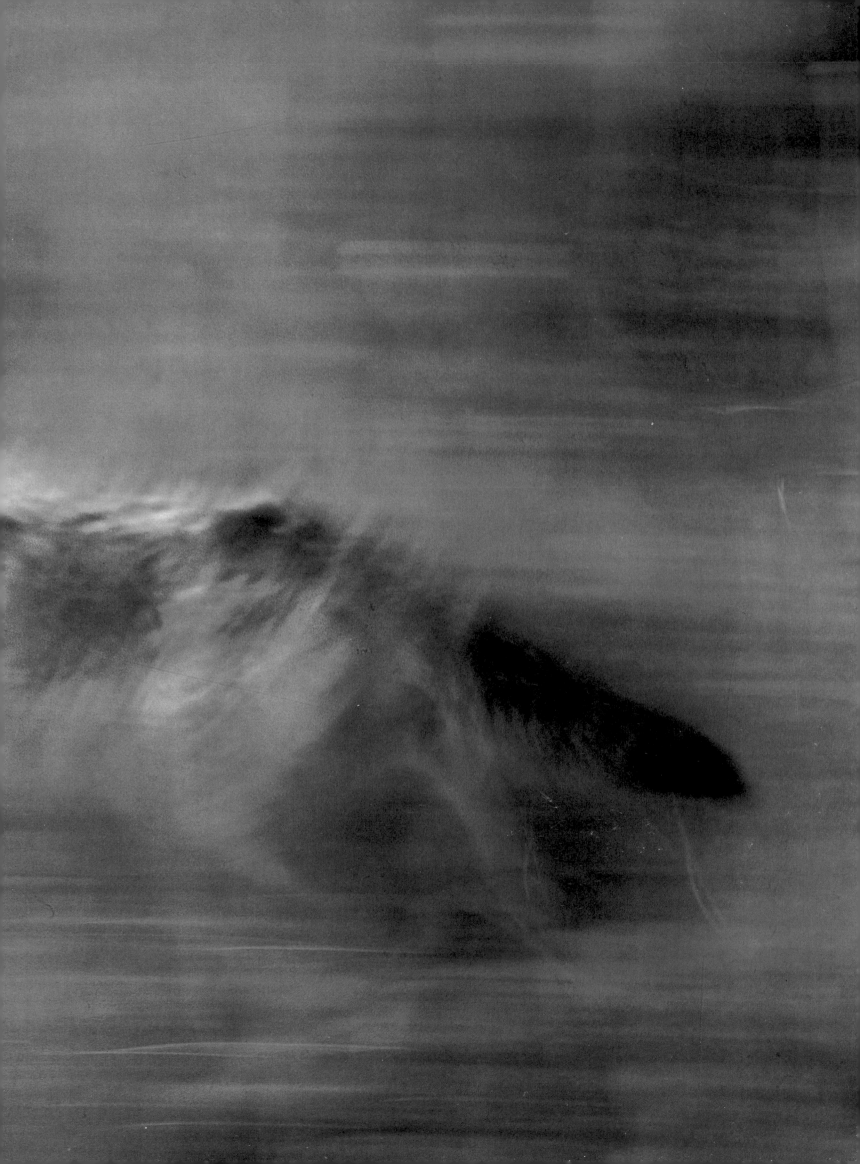

Photographic credits for the preceding illustrations:

The Hunters: Cats

Its passage heralded by the sharp yaps of the little muntjacs, or barking deer, a tiger walks the land. Moving fluidly, the massive muscles in its shoulders undulating to match its strides, it pads through a green tongue of woodland below a sunlit, grassy hillside. At the margin of the trees, the tiger pauses. The great, ruffed head turns slowly. The eyes, with chilling assurance, survey the landscape. Scattered in knots on the verdant slope ahead are sambar deer, large, big-eared creatures covered by coarse brown hair. They are feeding, their dark bodies gilded by the sun of the late afternoon. As the tiger scans the hillside, a wave of perturbation sweeps over the little clusters of deer. Heads are raised. Ears twitch. Nostrils quiver wetly, inquisitively, testing the soft breeze. The deer sense that, in the peace of the declining day, death has arrived.

The cats are many in species, diverse in size, but alike in mien and in that, from the largest, the 700-pound tiger *(Panthera tigris)* of Siberian snows, to the dainty margay *(Felis wiedii)* of South America, they are all killers. They live by killing, and the more efficiently they do it, the better the chances that their progeny will survive. It is the lot that has fallen to them in the natural scheme of things.

Adaptation to this end has conferred a terrible majesty upon some of them, especially the larger ones. It has made all of them sleek and silent, stealthy and supple, so that even the smallest seem to possess both an inscrutable cunning and vast dignity. It has armed them with muscles that can move fluidly but with sledgehammer power, with fearsome teeth and hooked claws that can either hold a victim fast or rake the life from it. The cats need all of these advantages, for they face the unending task of killing enough prey to survive, and must surmount enormous obstacles to accomplish it.

Concealment, then a sudden bound or explosive rush, is favored by most of the feline hunters, but within the overall pattern the hunt takes many forms. A stealthy, solitary stalk is the way of most cats, both the smaller ones, such as the African wild cat *(Felis lybica)*, and their larger relatives. The tiger creeps to within a few yards of the victim, then launches its attack from the rear or the flank, fastening its claws into the body of the prey while it bites the neck. It can be an ambush, favored by the small clouded leopard *(Neofelis nebulosa)* of southern Asia and the more common Afro-Asian spotted leopard *(Panthera pardus)* and the American bobcat *(Felis rufus)*, which may drop upon their prey from a tree. The hunt sometimes takes the shape of a spectacular acrobatic performance, like the leaps of the slender, long-legged serval *(Felis serval)* of Africa. With blinding speed, it bounds through the high savanna grasses, flushing birds, and springing after them to claw them out of the air six feet above the ground.

Whatever the style of the hunt, however, success depends upon the ability to work close enough to the prey in secret so it can be brought down in a single, decisive maneuver. The nature of the maneuver depends upon the particular physical assets of the cat that performs it. The cougar *(Felis concolor)* of the Americas is fast enough to overtake a deer on a course of a hundred yards or so. Usually, however, the conditions under which the cougar hunts are far from ideal, so it tries to get within a few yards where it can strike like an uncoiled steel spring, hurtling into its victim and even bowling it over. The caracal *(Felis caracal)*, a sand-colored lynx of Africa and Asia, relies far more on its speed. With hind limbs longer than the fore, it can run down even gazelles over a short distance. For these and the other cats, the hunt is almost always a solitary affair, except when mothers are teaching their young the skills of the killer. A major exception to the rule, however, is the lion *(Panthera leo)*, most social of all the cats, the only one to live in large family groups, and as often as not a team hunter. Lions, or more precisely lionesses, regularly hunt as a group, with some members of the pride driving prey into the jaws of the others.

The strength of the cats, relative to size, is awesome. A leopard, which seldom weighs more than a man, commonly hauls prey almost as heavy as itself into a tree and stows it in the branches for a later meal. Cougars, about the weight of leopards, have dragged off horses many times heavier than themselves. Buffalo, gaur, wild boar, oryx, moose, caribou, elk—all but the very largest herbivores regularly fall prey to the bigger cats. The prey is often quite varied, however, proof of the adaptability of the group. Caimans in the water and monkeys in the trees furnish meals for the stocky, powerful jaguar *(Panthera onca)*, whose range extends from the tropics into the southwestern United States. Tigers sometimes rove the mangrove swamps at seaside in search of fish and even turtles. Lions occasionally pounce upon rats, in imitation of their smaller, domestic relatives. Few of the cats, in fact, cannot afford to be generalists in dietary matters when the need arises. Carrion almost always is acceptable, and sometimes sought out—not only by the aged and the sick but occasionally even by the strong. Lions regularly drive hyenas away from prey which those much-maligned night prowlers have downed in a legitimate chase.

While the cats are adaptable predators, some of them are sufficiently specialized to demonstrate how intimate are the bonds between victim and killer. The Canadian lynx *(Felis lynx)*, for example, depends mainly upon the big varying hare, or "snowshoe rabbit," for food. The hares undergo a cyclic rise and fall in their populations. At the peak of the cycle, approximately every seven years, they swarm over the landscape, but a year later they have all but vanished. The size of the lynx population parallels that of the hares, but always a year behind. The intimacy of the lynx's ties with the hare is demonstrated by the cat's large, snowshoe feet, which mirror those that carry the hare over the crusted snow.

The effects of specialization also are revealed in the body of the cheetah *(Acinonyx jubatus)*, which relies mainly upon small, fleet gazelles as prey. The cheetah, unlike the other cats, is built not so much for stealth as for speed. Its conformation in some ways approaches that of the wild dogs, which run down their prey rather than surprise it. This large, spotted cat has exceptionally long legs, and its claws, like those of the dogs, are blunt. The claws of the cheetah do not serve as weapons, so there is no need to keep them sharp. Moreover, unlike the other cats, which can sheathe their talons and turn their paws into velvet gloves, the cheetah can withdraw its claws only partly. When it hunts, it relies on its keen eyes and capacity for a sudden burst of speed—up to 70 miles an hour—which no other creature on four legs can match over a short distance. The only hope of the cheetah's prey is to twist and turn until the fleet cat tires and, sensing its limitations, gives up the chase. Often, evasive tactics are successful, and the cheetah is left panting and hungry. But when the hunt of the cheetah goes in its favor, the pursuit ends in a roil of dust and a welling of blood. It is a stark reminder that the purpose of the hunting beast is the cessation of life, but for a reason—so that other life can continue. It is symbolic of the endless cycle of death and renewal that is nature.

145 and 146 overleaf. There are seven races of tigers scattered sparsely about the Asian forests, and all are considered to be endangered. Indeed, the Bali tiger is probably extinct, for it has not been seen on that island since 1952. Of some other races, less than a hundred animals may survive. Most numerous, if two thousand can be considered numerous, is the Bengal tiger (Panthera tigris tigris) of India. This solitary hunter is the largest cat on earth; a male may stand three feet at the shoulders, measure thirteen feet in length including its tail, and weigh 600 pounds. No animal is immune to the tiger's great strength —not even elephants, rhinoceroses, gaurs, or crocodiles—and a tiger may claim a kill from a leopard. Its appetite is voracious: a healthy tiger requires three tons of meat a year, the equivalent of thirty domestic cattle or seventy axis deer. But its meals do not come easy. To make a kill, a tiger must stalk to within a few feet of the victim, then bring it down with a final lunge. If it misses, the cat may chase the prey a hundred yards or so, but with little chance of catching it. Thus the tiger often augments its diet with more easily obtained fish, frogs, turtles, rodents, even locusts. Tigers readily take to water, swimming with ease across swift rivers, large lakes, or bays.

Named for the German naturalist Peter Simon Pallas, who was noted for his scientific explorations of Russia and Siberia in the eighteenth century, Pallas' cat (Felis manul) has small ears and long fur on its flanks, belly, and tail—adaptations for life in the cold, snowy steppes of central Asia.

Enemy of snakes in the tropical American rain forests, an ocelot (Felis pardalis) can dispatch even a large boa constrictor. Male and female ocelots hunt in cooperative fashion, meowing to one another like house cats, and they share in rearing the young.

Big, round ears that almost touch identify the serval (Felis serval) of Africa's woodland savannas. Resembling a scaled-down cheetah, the serval has long, slender forelegs that it uses to probe rocky crevices and dens in search of rodents.

149. *Cautiously stalking the dry grasslands, savannas, and bush country of Africa and southern Asia, the caracal (*Felis caracal*) preys on rodents, hares, ostriches, and small antelopes. It will leap into a low-flying flock of birds to bring down several at a time. This rare desert relative of the lynx once was trained to hunt game for man.*

149. *Hunter of northern forests around the world, the lynx (*Felis lynx*) has huge, hairy feet that enable it to travel over deep snow without sinking. In Canada, the life of the lynx is inexorably tied to the snowshoe hare, its chief prey; the hare population is cyclic, and when it plunges every seven years, lynx numbers likewise suffer a severe decrease.*

149. *The cougar (*Felis concolor*) once stalked the New World wilderness from the Atlantic to the Pacific, from northern Canada almost to the tip of South America. Hunted as a game trophy, persecuted as a killer of livestock, intolerant of civilization, the cougar has vanished from most of its historic haunts. Weighing 200 pounds, it can easily bring down a large deer and lug the carcass back to its den.*

150 *overleaf. Prey of the bobcat (*Felis rufus*) runs the gamut from little deer mice to adult deer. Growing to twenty-five pounds of muscle and sinew, the bobcat—its trademark is its tail, which appears to have been bobbed—hunts the forests, swamps, and deserts of North America from coast to coast, and from the Canadian border deep into Mexico. It exists surprisingly close to town and cities, but humans are rarely aware of its presence except during the mating season, late in the winter, when the bobcat rends the nighttime silence with frightening squalls and yowls. Conservationists fear for the future of the bobcat; since the luxury fur trade was denied the skins of endangered spotted cats, its beautiful pelt has come into great demand and this has triggered heavy trapping pressure.*

152 *second overleaf. Third largest of the world's cats, after the tiger and the lion, is the jaguar (*Panthera onca*), which ranges from Patagonia to the Mexico-United States border. Its name derives from an Indian word meaning "creature that overcomes its prey in a single bound," which tells much about the power of this 250-pound nemesis of deer, peccaries, and capybaras. Like the leopard, the jaguar has a melanistic color phase, but even in glossy black individuals the shadows of the spots are visible.*

154

154-155. *Lions* (Panthera leo) *are famous for the gentle and affectionate family life in a pride that may include two or three males, ten lionesses, and their young of varying ages. Male lions tolerate the rough romping and food stealing of the cubs, but there is a lot of noisy squabbling among adults over kills, and males often wrest food from their mates. After a meal, however, calm prevails. Lion cubs—two to four to a litter—weigh about three pounds at birth. They are weaned in three months and immediately begin taking hunting lessons from their mother. Young males must leave the pride at the age of three and a half years, but lionesses often remain for their entire lives. Wildebeests, zebras, and the ubiquitous little Thomson's gazelles are the usual prey of a pride.*

156 *overleaf and* **158** *second overleaf. The cheetah* (Acinonyx jubatus) *is capable of incredible speed—nearly 70 miles an hour—but only in short bursts. If it fails to bring down its carefully selected target within about 600 yards, the pursuit is given up, for by then the cat is totally exhausted. Cheetahs prey most often on medium-sized antelopes, but they can bring down an animal as large as a young wildebeest, killing with a throat bite—then first eating the liver, kidneys, and heart.*

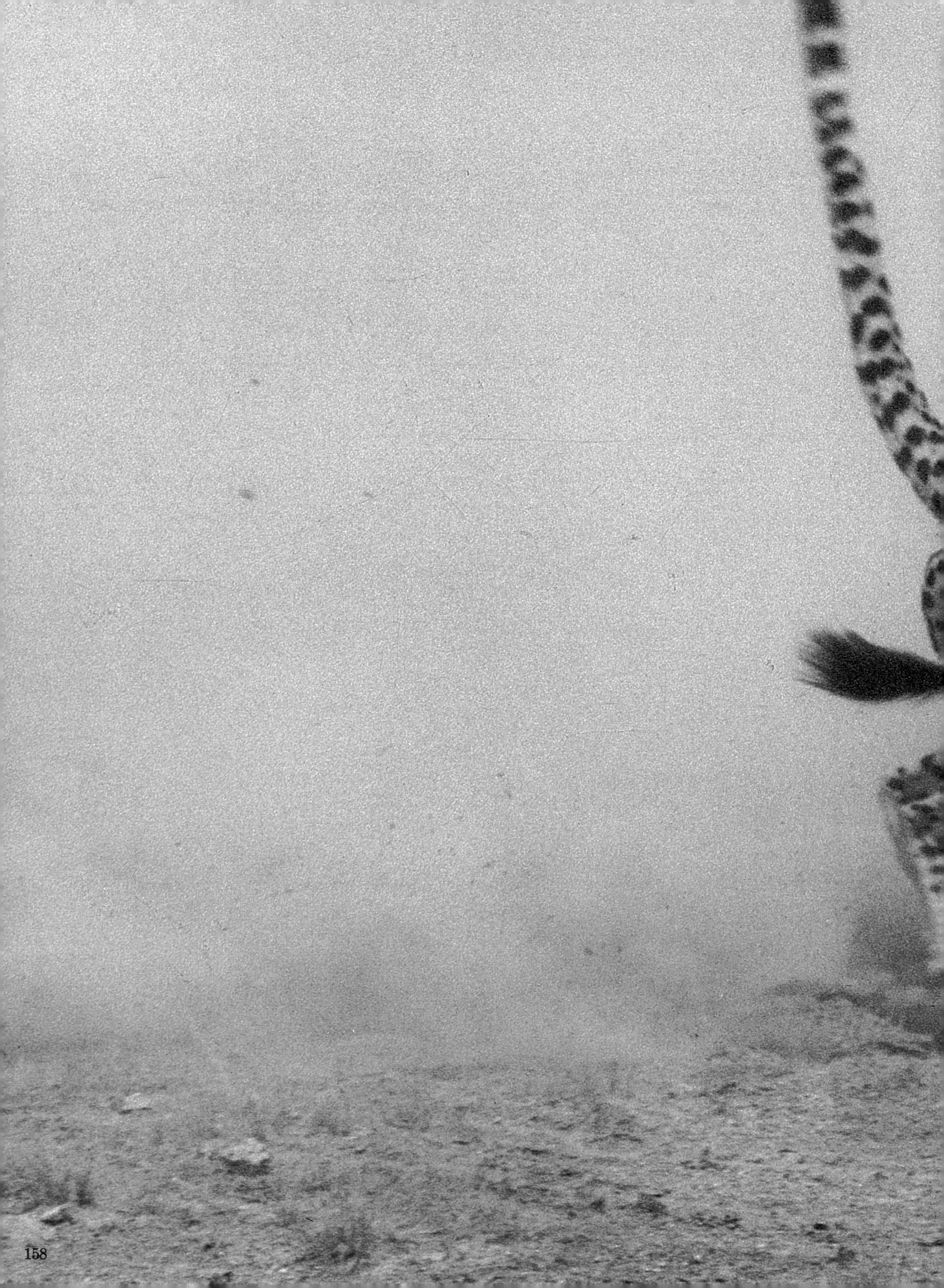

Photographic credits for the preceding illustrations:

Big Bears and Their Kin

The hulking form shambles to the water's edge, wades in, and plops down on its rump. Sitting in a stream with the water up to its chest, an Alaskan brown bear seems to be giving an impression of a big, hairy man enjoying a bath. There is much about the bears, in fact, that encourages anthropomorphism, for in many of their mannerisms, including even their plantigrade style of walking, with the entire underpart of the foot touching the ground, they seem to caricature humans. It is, however, an illusion that is quickly dispelled when a bear is aroused. A combat between two Alaskan brown bears *(Ursus arctos),* for example, is bone-chilling to witness, a battle between savage titans waged with the ultimate ferocity. Half-ton bodies no longer shuffle, but move explosively, almost lithely, with furious power. Piggy snouts quiver wetly and lips are drawn back to reveal huge canine fangs, yellowed and wickedly curved. Massive heads thrust forward or twist around to bury the fangs in matted brown hide, while the air resounds to thunderous snarls. Any anthropomorphic resemblances vanish in a scene of immense bestiality. The illusion, however, is persistent, and crops up again when an American black bear *(Ursus americanus)* stands up to beg for handouts in a national park or when a polar bear *(Ursus maritimus)* plunges with apparent joy from an ice cliff into the sea. The same sort of sham is carried on by certain members of another family of creatures, related to the bears and typified by the raccoon *(Procyon lotor).* In the raccoon's case, it is the creature's appealing manner and especially the use of its forepaw which contribute most to the fancy that it is really a little round man in a fur suit. Outside of the other primates, few creatures have digits which so closely simulate human fingers in form and function. The resemblance was not lost upon the Algonquin Indians. The name "raccoon" comes from the Algonquian word *aroughcoune*, which means "he who scratches with his hands." The "hands" of the raccoon are a marvel. It and its South American cousin the crab-eating raccoon *(P. cancrivorus)* have a wonderfully delicate touch, a sense so highly developed that they rely on it almost exclusively when dabbling in the shallows for crustaceans, fish, frogs, and other food.

Highly adaptable to human presence, the raccoon manages to exist even in cities, and thrives in suburbs, where the contents of backyard garbage cans provide an endless supply of food. Encountered while on a nocturnal foray into the trash can, a raccoon may pause with a purloined tidbit raised halfway to its mouth, and stare quizzically at the intruder as if to ask why it has been disturbed at its dining. At such times it can be a charming animal. If cornered by dogs, however, the raccoon displays another facet to its nature. It is transformed from a seemingly amiable clown into a ball of fighting fury. Cunning and tenacious, a raccoon is more than a match for a dog of its size—usually a dozen pounds but often much larger—as it crouches, comes in low, and aims for the throat. The raccoon's long, lean relatives the coatis *(Nasua)* can prove similarly difficult opponents for dogs, for they have especially long, sharp canine fangs which can slash through fur and flesh in one swift thrust.

The raids by raccoons on garbage cans testify to the omnivorous diet of these creatures, a trait shared by other members of its family and most bears. Coatis, ranging in troops of from a few to dozens of animals, poke their long snouts into virtually every crack and cranny in their path, eating everything from fruit to small mammals. The ringtail cats *(Bassariscus astutus)*, of Oregon to Central America, have a similarly varied diet, although they lean heavily on rodents and insects. Among the bears, the most specialized in terms of eating habits is the polar bear, for it spends most of its life on the ice or at sea, away from vegetation save an occasional strand of seaweed. The great white bear subsists almost entirely on flesh, mostly of seals, and has even been known to stalk humans on the ice in much the same way it hunts pinnipeds. Although the polar bear is the only one of its tribe that regularly kills large animals, most bears are capable of taking sizable prey. The Asiatic black bear *(Selenarctos thibetanus)*, a 300-pound creature with a reputation for aggressiveness, will occasionally attack and kill cattle and sheep. So at times will the big brown bears, including the grizzly of North America, which can down prey the size of a bison. Generally, however, they prey on creatures no larger than rodents. In midsummer, the Alaskan browns—some of which weigh more than 1600 pounds, stand almost a dozen feet high, and are the largest living land carnivores—gather at streams where salmon spawn. Shaggy coats dripping, the bears dash and plunge about in the foaming water, grabbing and snapping after the pink-fleshed fish. The brown bears of Eurasia, rather mild-mannered and only a third the size of the big Alaskan breed, consume more plant matter than their relatives, and, in fact, some of the European strain are almost completely vegetarian.

Termites are a staple of the sloth bear *(Melursus ursinus)* of the Indian subcontinent and Sri Lanka, and to get at the insects this dim-sighted creature has evolved a snout that approximates a vacuum cleaner. The lips of the sloth bear are hairless and flexible and can be protruded like a tube a considerable distance from the mouth. The bear can also close its nostrils whenever it wishes. After digging open a termite mound, the bear literally huffs and puffs to uncover and ingest its prey. Puckering its lips into a tube, with its nostrils shut to keep out dust, it blows away the loose dirt to expose the termites, then sucks them in, a process facilitated by a gap in its teeth resulting from the absence of two upper incisors.

Two other tropical bears, the spectacled bear *(Tremarctos ornatus)* of South America and the little Malayan sun bear *(Helarctos malayanus)*, obtain much of their food from the trees. The spectacled bear, so called because of the light markings that ring its eyes, hauls its 300-pound body aloft to a true nest, but forages much on the ground for palm stalks, new leaves, and seeds. The sun bear—the only bear smaller than an average human being—lives in a tree nest and eats fruit and the soft young growth of coconut palms. The raccoons and their relatives are even more adept in the branches. Raccoons often nest in tree hollows and take refuge in the trees when pursued by dogs and hunters. Troops of coatis readily scramble from the ground into the boughs. The ringtail cats are able climbers, generally living and denning on rock cliff faces. Two other members of the family, the olingo *(Bassaricyon gabbii)* and the kinkajou *(Potos flavus)*, are the equal of monkeys in the trees. The two species, both residents of the American tropics, regularly associate with one another. They are night rovers, wide of eye, and quietly search the forest canopy for the fruit they relish above all else. Rather gregarious—although it is not so much sociability as the presence of juicy fruit that brings them together—the kinkajous and olingos are long-bodied animals, with long tails as well. The tail of the kinkajou, an animal that moves with deliberation through the treetops, is the ultimate adaptation to the arboreal world, for, like the tails of New World monkeys, it is prehensile, serving as anchor and fifth hand. Looking like a rust-colored raccoon, the lesser panda *(Ailurus fulgens)* of southwestern China and bordering lands further demonstrates the arboreal proclivities of the raccoon family. The little red panda sleeps in the trees by day, its brushy tail wrapped around its head like the tail of a sled dog in the snow, and although it generally forages on the ground, it uses the trees as a refuge and tree hollows as dens. Even the giant panda *(Ailurpoda melanoleuca)*, as much a model for children's dolls and figurines as the "teddy bear" koala of Australia, will climb into the trees if pursued, although it can weigh upward of 300 pounds. Few mammals have caused so much difficulty for those people whose job it is to classify the members of the animal kingdom, assigning them to neat little groups. The giant panda is a source of puzzlement because it shares characteristics with more than one group. Taxonomists seem unable to agree whether it is a bear, a member of the raccoon family, or a bridge between the two groups. All they are sure of is that it is one of the three. The black-and-white panda has been a problem for the classifiers for only a century or so, because the world of Western science did not know of the creature until the late 1860s, when Jean Pierre Armand David, a Roman Catholic priest and naturalist, ventured into its rugged bamboo-covered territory in western China and Tibet and in all likelihood became the first European to see the creature. For the peoples of the West, David had found a new and curious animal. For the classifiers, he had created a problem that more than likely cannot be solved.

165 and 166 overleaf. The grizzly or brown bear (Ursus arctos) has been the center of many controversies ever since white man first intruded into its domain in the North American West. Most have been serious matters of survival for both bear and man. Not so the long-running battle over its name—or names. As one noted zoologist wrote, "Probably no other piece of research has brought dignified mammalogists nearer to name-calling and nose-punching than the question of correctly classifying the grizzly bear." The argument is between the "lumpers" and "splitters" in the taxonomic fraternity, which relies heavily on the shape of mammals' skulls and teeth in determining their classification. And the problem is the great variation in skulls and teeth among these big carnivores. Thus one scientist, a splitter in the extreme, decided there were eighty-four different species and subspecies of grizzly and brown bears in North America, and he claimed five distinct species could be found on one Alaskan island only a hundred miles long! Today, however, the lumpers rule. They tell us only a single species of brown bear is found across the entire Northern Hemisphere. Smallest of the local races is the 200-pound Syrian bear; largest of all is the Kodiak bear of coastal Alaska, a giant that stands nine feet tall on its hind legs and weighs 1600 pounds. All have that same ominous appearance—a hollowed-out face and a big hump between the shoulders. In North America, the term "grizzly" is given to the bears of the Rocky Mountain wilderness and inland Alaska, "brown bear" to those that haunt the Alaska coast, feeding on grass, berries, and other vegetable matter much of the year but growing fat in midsummer on the salmon that pack rivers and creeks on spawning runs from the sea.

168. *Although the giant panda* (Ailuropoda melanoleuca) *is a symbol of wildlife conservation around the world, its life in the mountain forests of China remains a mystery more than a century after its discovery. Scientists know that the giant panda eats copious amounts of bamboo shoots, a food with low nutritional value, and that its striking black-and-white coloration is an effective camouflage as the 300-pound animal sits on its haunches high in a tree. But what little else is known about panda behavior has been gained from studies of the very few animals in zoos. The panda is considered a national treasure by the Chinese, who call it* beishiung-chin, *meaning "white bear." Taxonomists still have not decided the giant panda's true relationships, some saying it is a raccoon-like bear, and others that it is a bear-like raccoon.*

169. *A family group of lesser pandas* (Ailurus fulgens), *long-tailed Asian members of the raccoon family. The lesser panda frequents the mountain forests and bamboo thickets on the south-facing slopes of the Himalayas, from Nepal to Burma and the Chinese provinces of Yunnan and Szechwan. It too feeds extensively on bamboo, grasping the shoots with prehensile thumbs and chewing the tough fiber with powerful jaws and large teeth. But unlike the giant panda, it is not strictly vegetarian, for it eats nestling birds, eggs, rodents, and insects captured on its nocturnal ventures.*

170 *overleaf. Stalking the pack ice and frigid waters around the North Pole in search of seals and walruses, the polar bear* (Ursus maritimus) *is one of the great nomads of the mammal kingdom. It spends most of its life at sea, often floating hundreds of miles from land on great ice floes. A powerful swimmer, the polar bear has front paws that are partially webbed, and the soles of its broad feet are furred for insulation against the cold and to provide traction. Polar bears come ashore only occasionally, varying their diet with tundra berries or hunting down a caribou or musk-ox. In winter, females make a snow den on land to give birth to their cubs, which number from one to four.*

Photographic credits for the preceding illustrations:

The Deer Tribe

Lying in the snows of January, an antler discarded by a
whitetail buck affirms the continuity of nature, the
marvelous cycle of life that progresses unbroken through
the endless round of seasons. Crusted with ice, perhaps,
or gnawed by rodents, the jettisoned tree of bone on
the woodland floor promises new life to come from old,
but at the same time cautions that certain variations
in the way the pledge is fulfilled may elude human
understanding. Antlers are the peculiar property of the
deer family, half a hundred species of generally graceful
cud chewers native to the Americas, Eurasia, some
large Far Eastern islands, and northwestern Africa, and
widely introduced beyond those areas. Although sometimes used as a weapon, the primary role of antlers
is sexual, because ultimately they expedite the inheri-
tance by the young of genes rich in survival value.
The sexual import of antlers is signified by the fact that
they are carried only by the males and employed in
bruising tournaments to gain dominance and the right
to mate with females. There are exceptions to the all-
male rule. Both sexes of the Eurasian reindeer (*Rangifer
tarandus*) and its North American counterpart, the
caribou, carry antlers. However, antlers are entirely
missing from either sex of two Asian species, the
Chinese water deer (*Hydropotes inermis*), which has
been introduced into England, and the small musk deer
(*Moschus moschiferus*). Instead, the two Asian deer have
sharp canine tusks in the upper jaw. Oddities today,
they may resemble the creatures from which the deer
tribe arose, for the ancestors of the deer showed no
sign of antlers. During the past 25 million years or so,

however, antlers have appeared and evolved into myriad sizes and configurations. The smallest deer, the foot-high pudus *(Pudu)* of South America, and the largest, the hulking, 1800-pound moose *(Alces alces)* of North America and Eurasia, also stand at opposite ends of size when it comes to antlers. Those of the pudus are slender, finger-length spikes. The moose carries massive, palmate antlers that have a spread of more than six feet and outweigh the entire body of the twenty-pound pudu by a factor of five. The giant stag, or "Irish elk," which strode across the open landscape of Europe during the last ice age, had antlers that spanned nine feet and weighed about 150 pounds.

Changes in the antlers as they bud, grow, and eventually deteriorate mirror changes in the life of the deer as they move through their yearly reproductive cycle. Shedding of the antlers signals that mating is past, that males may join one another or on occasion even mix with the females without conflict or sexual involvement. The Eurasian roe deer *(Capreolus capreolus)* mates in summer and sheds in autumn. The Père David's deer *(Elaphurus davidianus)* of China, extinct as a wild animal for centuries but preserved in zoos, follows a similar cycle. The American whitetail *(Odocoileus virginianus)*, which ranges in size from more than 300 pounds in the North to a 50-pound race of midgets in the Florida Keys, mates in autumn and is without its antlers by midwinter. The gorgeously spotted axis deer *(Axis axis)* of India and Sri Lanka usually mates in the spring, and when it does, drops its antlers in August. But axis deer have also been known to mate at any time of the year, and shed in corresponding fashion, a situation rather typical of tropical deer. For example, the big swamp deer *(Blastocerus dichotomus)* and the pampas deer *(Ozotoceras campestris)* of South America lose their antlers at odd times of the year, and coincidentally have no regular breeding season. Nor do the males of these species engage in rutting battles. Reportedly, the same is true of stags of the sleek Asian barasingha *(Cervus duvaucelli);* their antlers are still encased in a coat of velvet at breeding time.

For most deer, however, the rutting season opens just as the antlers become gleaming hard. Their maturation heralds the beginning of competition between the males, a contest that may be waged at levels of which we are oblivious. The antlers almost certainly express the dominance of the fittest males in more subtle ways. A splendid set of antlers unquestionably advertises the vitality of its owner, and indeed, the growth of antlers is intimately linked to the production of the male hormone testosterone. A stag that is neutered while it is immature, and thus deprived of testosterone, never produces antlers. A spayed female injected with it develops a rudimentary set of antlers.

Mature antler is bone which has grown rapidly out of the skull and then died. No other bone grows so profusely after birth, and the accelerated buildup of cells as the antler develops has been likened to the runaway multiplication of bone cancer cells, but with the control and direction needed to form a precise structure. While growing, the antler is linked internally to the blood supply of the skull. As the bone hardens, the linkage is carried out through vessels in the velvet, the tender skin over the growing antler. Eventually, as hormones ebb and flow in the body, the blood supply dwindles, and the velvet shrivels, frays, and is rubbed off on trees and other objects.

When their antlers are polished and ready, the males, which generally have lived by themselves for several months, roam in search of females. Most male deer announce their readiness to mate—and to do battle— with challenging calls that blast over the landscape. The moose emits a raspy roar. The North American mule deer (*Odocoileus hemionus*) gives out loud grunts. The graceful fallow deer (*Dama dama*), introduced to Europe from the Middle East as early as classical times, coughs. The sika deer (*Cervus nippon*) of Asia utters shrill, piercing sounds. The most explosive calls of all, however, come from the European red deer (*C. elaphus*), and especially its North American relative the elk (*C. elaphus*).

The red deer, fabled stag of medieval legend and lore, and of the golden artwork of the Scythians, challenges with a harsh, grating bellow that leaves little doubt it means business. The bugle of the elk is one of the most thrilling sounds in the animal kingdom, as during September and October the valleys of the Rocky Mountains and a few other havens left to the creatures echo to their calls. Muzzles lifted to the sky, antlers tilted back on necks swollen with blood, the bulls scream clarion clear. The mere bugle of a big bull can send a rival fleeing. Elk and, in fact, other deer in rut can be lethal creatures as they walk stiff-legged in search of competitors.

There comes a time, however, when males of equivalent size and strength clash. Elk and moose pairs come together with head-shattering clashes. Whitetails and mule deer fence close up, pushing and shoving. Sometimes one is killed or injured, but usually the weaker backs off. One of the few deer that seemingly fights to kill in rutting bouts is the hog deer (Axis porcinus) of Asia, which charges with antlers canted to the side so the points reach the opponent's body rather than engage his tines. Once the branching structures of deer antlers engage those of an opponent, the sharp points of the tines usually are kept at a distance. The greatest danger to the combatants is that the antlers will lock, leaving them to slow, grim death by starvation.

The antlers are seldom used for fighting predators, and for that matter are either absent or absolutely useless most of the year. Most deer box enemies with their razor-sharp hooves. Whitetail deer kill snakes by stamping upon them. A moose can break the skull of a wolf or a bear with a swipe of its huge front hooves. Male deer that triumph in mating battles often gather large harems about them. Caribou and reindeer bulls may have up to forty cows, some with their young. Mule deer are somewhat lackadaisical, and gather only three or four females, which are allowed to wander off. Moose often remain with one cow, but also sometimes spread their attention among several. The roe deer buck fixes his desire on only one female, which he pursues in long chases over the countryside. Often the pursuit takes the shape of a small circle, which the deer tread into fields and meadows. The imprint of their revels is the so-called witch circle of European lore. Once mating is complete, male deer leave the females. Often the males remain in bachelor herds until the next rut approaches, but sometimes, especially in winter, deer will form loose herds of both sexes.

With the end of breeding, the antler loses its purpose. Ironically, it is cast off because it starts to grow again. The spurt of new growth occurs on the frontal bones of the skull, in the bony platform from which the antler arose. The antler, however, is dead, and cannot respond to growth, so the force of new cells piling up below pushes it up and away from the skull, until it is so loosely joined that a casual blow, or even a vigorous shake of the head, will dislodge it. Fallen to the ground, the antler is a silent message declaring that nature regenerates itself even as it dies.

177. *Late-afternoon sun highlights the budding, velvet-covered antlers and outsized ears of a mule deer* (Odocoileus hemionus), *the deer of mountains and deserts of western North America. Mule deer are noted for the strange way they run in flight, bounding in four-foot-high leaps, looking backward each time to check on their pursuer. In some areas, local people call it the jumping deer. Mule deer are plagued in summer with ticks—thousands of them on a single animal—and, like the oxpeckers that accompany the wild buffalo of Africa, magpies and jays will perch on a deer's back and pluck off the pests. Occasionally one deer will chew at the parasites on another deer. Black bears and the rare grizzly prey heavily on mule deer fawns, and coyotes will try to bring down an adult, but more often than not it will be routed or killed by flailing hooves. The mule deer's big eyes give it superb vision in the dim light of dawn and dusk.*

178. *A month after its birth in late June, a mule deer fawn can keep pace with its mother as she moves about their home range of about a square mile. Though the fawn is still nursing, it begins to sample the mule deer's varied diet of foliage of all kinds of mountain shrubs and trees and abundant summer mushrooms. By the time the fawn is weaned in September, its spotted coat has been replaced by one of long, heavy hair. This shaggy winter coat is noted only on first-year deer, who are facing the most critical season of their lives.*

179 *top. Except for its long antlers and a bushy tail, the hog deer (Axis porcinus) of India and Southeast Asia indeed does resemble a pig: it is squat, stocky, and has a piglike gait when it runs across the grasslands.*

179 *below. Lacking antlers, the male musk deer (Moschus moschiferus) instead has canine tusks that jut three inches below its jaws. This small, timid deer of central and northeastern Asia is relentlessly hunted for the musk gland of the male, because the small quantity of musk it contains is in great demand for perfumes and soaps. Countless females and young also perish in traps set by native hunters.*

180 *overleaf. The spotted red coat of the axis deer or chital (Axis axis) adds color and beauty to the grasslands of India and Sri Lanka. Of all Asiatic deer, this is the most gregarious, forming herds of males, females, and young that number more than one hundred.*

182. *Antlers of the sambar (*Cervus unicolor*) of southern Asia uniformly have only six tines, a fact noted by Alexander the Great on his campaign in northern India in 326* B.C. *and subsequently recorded by the Greek philosopher and zoologist Aristotle. Thus this large but stealthy deer of bamboo jungles and dense forests is sometimes called Aristotle's deer.*
183. *New antlers are growing on an axis deer. Like all deer of tropical regions, axis stags shed their antlers irregularly, during all months of the year. Antlerless stags form all-male herds, an individual returning to the mixed herd when a replacement set is fully formed.*

184 *overleaf. A flashing white "flag" disappearing into the dense forest is often a hiker's only clue that he has startled a white-tailed deer (*Odocoileus virginianus*) into flight. The large, waving tail is a signal that helps to keep small groups of this common New World deer together in heavy cover.*
186 *second overleaf. During the rut of autumn, as dusk approaches in the national parks of the Rocky Mountains, American elk or wapiti (*Cervus elaphus*) leave the forests for grassy meadows. There, announcing his claim to a harem of cows, a bull elk will bugle a long, echoing, three-note call. If he is challenged by a rival, a violent battle is soon to erupt, the two bulls charging from thirty paces and colliding with* incredible force. On rare occasions the antlers of two bulls will become inextricably locked together; the combatants are then doomed to death, and the cows over which they fought will be ruled by a lesser bull who was an onlooker at the fatal fight.

190 *overleaf. It is autumn on the Alaska tundra, and shreds of velvet hang from the tender, blood-red antlers of a bull caribou* (Rangifer tarandus). *The caribou herds have formed up for the migration to their winter range, and a large bull like this will have stored up fifty pounds of fat on its back and rump in preparation for the forthcoming battles of the rutting season. Scraped clean and polished against willows and spruce, the antlers will be put to hard use before they are shed as the heavy snows fall.*

188-189. *The world's largest deer, the moose* (Alces alces) *is a giant resident of marshy forests around the Northern Hemisphere, browsing willows and poplars or wading into ponds and streams to feast on aquatic vegetation, submerging its head—and sometimes its entire massive body—to reach roots and stems. A bull moose, with flat, palmate antlers that spread seventy inches, can weigh 1800 pounds.*

Grazing Herds

Hours before the fall of night, the savanna has been darkened, not by shadows but by the slate-colored forms of the wildebeest, immense herds of them arriving from the dry lands to the south. Leaving the dusty expanses of the Serengeti Plain to the gazelles and ostriches, the wildebeest have made their annual flight from drought. When night finally does cloak the landscape, the air is heavy with the presence of the herds, with a great but muffled stirring of bodies, punctuated by breathy snorts and occasional braying.

The migration of the Serengeti wildebeest *(Connochaetes taurinus)* echoes the mass movements of the enormous herds of hoofed herbivores that once occurred throughout the vast plains of Africa, Eurasia, and North America. These open grasslands have for millennia been the home of multitudes of grazing or browsing creatures, chiefly antelopes, wild cattle, and their relatives. Some of these animals, such as the wisent *(Bison bonasus)* of Europe and the bushbuck *(Tragelaphus scriptus)*, live in the forests, but for millennia the vast herds have inhabited places of far horizons. Today, many of the herds have vanished, or are irreversibly diminished, and few range as freely as in the past. Even so, there are places such as the Serengeti where the herding beasts live much as before, their existence governed by an endless round of different living patterns, repetitive but transitory, distinctly observable but meaningless when separated from the cycle.

For the wildebeest—and until a century ago for the American bison *(Bison bison)*—one of the patterns of life is a seasonal change in range, accomplished by a migratory trek. But for all the horned herds, the cycle of nature's year causes changes that shape their lives even more profoundly than a seasonal shift of scene. Entire animal societies, unshakably stable part of the year, break down and are restructured. The behavior of individuals may change so drastically that they act like entirely new forms of animals. For instance, the male Grant's gazelle *(Gazella granti)* or the hartebeest *(Alcelaphus buselaphus)* that has lived peaceably with others of its sex for months suddenly becomes solitary and cantankerous. The bull African buffalo *(Syncerus caffer)* which has lived on the fringes of the herd becomes gregarious. The female greater kudu *(Tragelaphus strepsiceros)* which has discouraged the approach of bulls with solid butts changes tack and allows a big bull to sidle up alongside her. The timing of such changes of pattern is geared to the survival of the species and of individuals. Correlated to conditions in the environment, the timetable promotes breeding by the adults that are the most fit, the birth of the young when life is easiest, and maximum use of food when it is scarce.

The key to the entire cycle is often the territorial behavior of the males. Among the herds, it varies considerably among species, and even individuals, in the amount of territory claimed and how long it is held, and even whether or not they hold territory. Generally, however, territoriality is strongest when the land is lush, weakest during periods when animals are hard-pressed just to stay alive.

This is dramatically demonstrated by the impala *(Aepyceros melampus)*, a medium-sized African antelope. At the beginning of the rainy season, when pasturage is rich, the dominant males stake out their territories, about 500 square yards each. The claims are marked with urine, feces, and an oily substance secreted by glands and rubbed onto bushes. The territorial males are easily picked out by the way they stand on their ground—tall and erect, as befits rulers. For up to three months, each male holds his territory against other adult males that seek to take it. Roaring fiercely to proclaim his dominance, the ruler of a particular chunk of territory is kept busy rushing back and forth to check on the small bands of young, non-territorial bachelor males that sometimes wander through his area, and driving off serious challengers. He may dash after an interloper, engage in a fencing match, using his lyre-shaped horns, then turn to meet the next challenger. The loser of the contest ceases territorial behavior and joins the bachelor band. Before becoming a challenger for territory again he must work his way up through the ranks of dominance within the bachelor herd.

Once a male has established a territory, and as long as he holds it, he has the sole right to mate with the groups of females that wander into it in quest of food. When the territorial male is not trying to ward off rivals, he is intent on holding the females in his space and mating with those that are receptive. The females are rather fickle, and males that have the best pasturage in their territories retain the females the longest. An ample supply of food near at hand is also a tremendous advantage for the male as well, because between defending his land and winning the favor of females he has little time to feed. By the end of his tenure his condition deteriorates and he is vulnerable to dispossession by males that were no match for him earlier. The withering of the food supply as the dry season approaches sends the males farther and farther away from their little kingdoms in search of something to eat. Boundaries between the claims are crossed with increasing frequency as the food dwindles and self-preservation replaces the reproductive urges. The dominant males lose interest in maintaining their realms. Eventually, especially if the drought is severe, all defense is abandoned and the territorial system evaporates, as though carried away by the wind that whistles dryly through the acacia thorns. No longer are the various sexual groups of impala kept apart by the domineering rulers of the territories. The antelopes—females, bachelors, and the deposed rulers—mix and focus their energies upon the desperate task of finding food and water on the sun-baked plains.

For most of the herding, hoofed multitudes, the onset of hostile environmental conditions is followed by the disappearance of territorial behavior, a lowering of aggression, and the urge to gather with others of their kind. It happens to the huge gaurs *(Bos gaurus)* of southern Asia during particularly severe droughts, and to the African buffaloes, which gather on the sides of hills and valleys when the dust devils swirl over the landscape. When the rains cease in East Africa, the Thomson's gazelles *(Gazella thomsoni)*, which have been fiercely territorial, herd and migrate from the open plains to the bush in search of fresh pastures and water. On the steppes of Eurasia, the coming of winter ends the vicious territorial combats of the male saiga antelopes *(Saiga tatarica)*. The bulbous-nosed saigas merge into immense herds that run before the wind when blizzards sweep down from Siberia.

Only with the disappearance of territoriality can creatures such as the saigas form great, mixed herds. When the wildebeests migrate, for instance, males inspired by the territorial urge leave the main herds for a day or so, establish their transitory rule, mate with whatever females the can corral, and then return to the main body of antelopes. In contrast, the plains zebra *(Equus burchelli)*, which often travels with the wildebeest and is dependent upon the same pastures, maintains a form of society that does not change with shifting environmental conditions. For the zebra, survival is best served by living year round in small family groups dominated by a king stallion, although stallions also live together in bachelor groups.

For the creatures such as the impala and hartebeest, the cyclic phenomenon of territoriality assures that the bulk of the young will be sired by the finest of the males and that the offspring will be born at a time when their chances of survival are highest. This is accomplished by a marvelous natural synchronization between the time when the dominant males mate, the length of the gestation period, and the onset of the rains, when the land is lush and productive. It is all timed so that the young conceived at the height of the mating season will appear when there is plenty of pasturage, first for their mothers, who consequently can supply sufficient milk, and later for the newly weaned offspring. Moreover, having large numbers of young born at the same time guarantees that a sizable proportion will survive the ravages of predators, which, after all, kill just enough to satisfy their own needs. The young born late, however, have considerably less chance of surviving, not only because of a depleted food supply but also because the predators, no longer luxuriating in an overabundance of prey, pick them off one at a time.

Brutally efficient, the system nevertheless works out for the best interests of all species concerned, for the latecomers among the young are often the offspring of less fit males, which have managed to mate at the end of the territorial season, by chance or because the dominant males are through.

197 *and* **198** *overleaf. Whether a solitary bull sharply outlined by the morning sun on the South Dakota prairie, or a herd huddled against pelting snow and subzero cold, the American bison* (Bison bison) *is a living legend—and a reminder of how close we came to a great tragedy. Once there were 50 million bison on the Great Plains; by 1889, when the market hunters gave up their notorious slaughter, only 541 animals remained! An aroused public succeeded in protecting the survivors, and today there are 25,000 American bison roaming semi-free in the large national parks and wildlife refuges that preserve parts of their historic range. This is the largest land mammal in the New World; a bull, standing five feet at the shoulders, can weigh 2000 pounds. But the cow, although smaller, is the leader of a family group. A cow will nurse its single calf for an entire year, and the young bison will remain with its mother until it is sexually mature at the age of three years.*

200 *second overleaf. Sharing the North American plains with the bison when the white man pushed into the western frontier were an estimated 40 million pronghorns* (Antilocapra americana). *They were not in competition for forage: bison graze the grasses, pronghorns browse shrubs and weeds. Nor did they deplete their pastures, for both species are nomads. There is no faster animal in the New World than the pronghorn: it can maintain a cruising speed of 30 miles an hour over several miles, with bursts up to 40 miles an hour. But it is unable to jump man's barbed-wire fences, and when the wild grasslands were claimed for cattle, the pronghorn population plummeted, reaching a low of 30,000 in the 1920s. Conservation measures have restored pronghorn numbers somewhat, but even so they amount to only one percent of the size of the historic herds. The pronghorn's horn is unique in the mammal kingdom: the bony permanent core is covered with a hard sheath of fused hairs that is shed annually and consumed by rodents.*

202. *The national symbol of South Africa, the springbuck* (Antidorcas marsupialis) *is named for the peculiar way it jumps, or stots, when excited. Holding its legs stiff and its hooves together, the springbuck bounces vertically like a child on a pogo stick, a pouch of white hairs opening on its arched back and flashing at the peak of its ten-foot leaps. One "pronking" springbuck is likely to start an entire herd bouncing. Herds of hundreds of thousands of springbucks once migrated across the dry grasslands, but they were decimated late in the nineteenth century by colonists attempting to protect their crops and by a great rinderpest epidemic in 1896.*

203. *The word "gazelle" is synonymous with grace, delicacy, and beauty, and Grant's gazelle* (Gazella granti), *seen in large numbers on the open plains of East Africa in the company of herds of zebras and wildebeests, is no exception. Male gazelles of varying ages form bachelor herds, but individuals constantly challenge the domination of the buck overseeing a nearby herd of females and young. Grant's gazelles are a common prey of lions, cheetahs, wild dogs, and hyenas.*

204 *overleaf. A water hole in Natal reflects the beauty of a group of female nyalas* (Tragelaphus angasi). *In strange contrast to the hornless female, the bull nyala has twisted horns and a shaggy blue-black coat with a ridge of white hairs down the center of its back. One of the loveliest of all antelopes, the nyala is restricted to the almost impenetrable lowland brush country of southeastern Africa, almost entirely in Mozambique.*

has a circular territory 60 to 200 feet across. With whistling, stamping, and posturing, the male kob displays to and mates with any female that enters his particular field.

208 *overleaf. Armed with scimitar-shaped horns that may exceed five feet in length, the sable antelope* (Hippotragus niger) *fears not even a circling pride of lions. Herds of up to eighty sables — rust-colored females, their young, and one jet-black old bull — roam the woodland savannas of southern Africa.*

210 *second overleaf. Weighing 1800 pounds and carrying five-foot horns, the African buffalo* (Syncerus caffer) *has a reputation — at least partly deserved — as the continent's most dangerous animal. Although it is normally peaceable, a wounded buffalo will lie in ambush for a hunter, and an old bull may stalk and charge a man without provocation. Ox-peckers are constantly in attendance on these wild cattle, ridding them of the torment of blood-sucking ticks.*

212 *third overleaf. A herd of blue wildebeests* (Connochaetes taurinus) *in flight across the Serengeti Plain of Tanzania. Widely dispersed during the rainy months, wildebeests gather in tremendous assemblages of hundreds of thousands in the dry season, moving forever about, single-file, seeking water and pasture. The wildebeest is a favorite prey of lions, hyenas, and wild dogs, and predators may claim eight out of every ten calves before they reach maturity.*

214 *fourth overleaf. A zebra cannot be mistaken for any other creature, but within the three species of zebras — indeed, within the most widely distributed species, the plains zebra* (Equus burchelli) *— there is great variation in the striping. Zebras find a certain amount of safety from their chief nemesis, the lion, by mingling with large herds of other grazing animals — oryx, wildebeests, even ostriches. Their primary defense in an attack is speed — up to 40 miles an hour — but their hooves and teeth are weapons to make a lion wary.*

206. *Both male and female oryx* (Oryx gazella) *sport ringed, rapier-like horns, and those of the female are the longer, jutting up to four feet. If attacked by a lion, the oryx lowers its head and directs its menacing weapons forward; but in rutting season, fighting bulls only butt their heads, their horns carefully arranged to avoid injury. So sharp are the horns of this large, beautifully marked antelope — scattered about Africa and Arabia in several subspecies — that natives use the tips as spear points.*

207. *The kob* (Kobus kob) *inhabits open grasslands of central Africa near swamps and rivers into which it can retreat to escape midday heat. Herds of kobs establish breeding arenas, within which each male*

The Domesticated Ones

On the Mediterranean coast of France, just west of
Marseille, the two forks of the Rhone River flow to the
sea in a welter of marshes, mudflats, and shallow
lagoons. This vast delta region, known as the Camargue,
is a land of beauty, strange and almost alien in aspect.
Wild boars slosh through the shallows. Amphibians and
reptiles swarm in freshwater marshes. Foxes skirt the
salt lagoons where—astonishingly—vast flocks of greater
flamingos congregate. And over this curious landscape
roam herds of fierce black cattle and wild free horses,
many almost white as snow.

The horses and cattle of the Camargue are domestic
animals, but over long years they have been allowed to
roam relatively free of interference, although the
cattle, mostly privately owned, are branded and the
horses are sometimes caught and broken as mounts for
the herdsmen. For all purposes, the cattle and horses
of the Camargue are feral, reverted to an untamed state
and living much as their wild ancestors did in pre-
historic times. They are a reminder of the close links
between many domestic creatures and their wild fore-
bears and of how recently, in the long course of
mammalian evolution, man has shaped domestic breeds.
The domestic horse, for instance, was bred from the wild
horse *(Equus caballus)* which ranged the steppes of the
Ukraine more than 5000 years ago. Domestic breeds
undoubtedly were developed from a number of local,
and now extinct, races. The only wild horses alive today
are the zebras and the Mongolian horse *(E. c. przewalskii)*.
The latter barely survives in the wild, and there is considerable thought that it may be a completely distinct

species from that which produced domestic horses. The horses of the Camargue, like the mustangs of the American West, are not wild in the strict sense, but actually are a domestic strain that has run free for 2000 years.

In the course of domestication, the traits that have made for manageability have been accentuated, and those that pose inconvenience or downright danger to humans have been eliminated by selective breeding. Thus it is that the male domestic yak *(Bos grunniens)* is half the size of its rare wild relative, which weighs more than a thousand pounds. Similarly, most donkeys are smaller than the pony-sized North African wild ass *(E. asinus)*, their graceful, fleet ancestor, and many strains of domestic water buffalo *(Bubalus bubalis)* are considerably less bulky than the wild variety, and have less imposing horns, sometimes none at all.

The water buffalo has been harnessed for use because of its adaptability to a very special set of environmental circumstances. It can be used for milk production, and more importantly as a beast of burden, in the hot, swampy places that are all but unlivable for domestic cattle and oxen. Because the water buffalo can work in the mud and water of rice paddies, marshes, and jungles, it has gained in numbers to perhaps 150 million animals worldwide. Water buffaloes have been spread to such diverse lands as Italy, Brazil, and Australia, running feral in parts of the latter two countries.

The same adaptability to quite special surroundings characterizes several other domestic ungulates. The yak, the llama *(Lama glama)*, and the alpaca *(L. pacos)* are all suited for the windswept reaches of the high mountains. All have been developed from beasts that evolved on arid tableland up to altitudes of 17,000 feet, and even higher in the case of the yak. The llama, a beast of burden, and the alpaca, kept for its long, fine wool, descend from the far-ranging guanaco *(L. guanicoe)*, which while it roams to the edge of the sea—and even to some islands—is very much a creature of the high, arid plains of the Andes. The yak's wild relative inhabits the wind-whipped deserts of northern Tibet, a bleak, barren wilderness. The country inhabited by these animals, and in which their domestic relatives are employed, represents some of the most difficult, dangerous terrain on earth, where surefootedness is at a premium and a capacity for subsisting on tough, sparse vegetation is essential. The shaggy winter coats that serve the guanaco and wild yak so well under near-glacial conditions have become in their domesticated relatives a source of extremely useful wool. Selective breeding, for example, has produced alpacas with fringes of wool so long they nearly brush the ground; the long, thin wool is considered to be among the world's finest and is extremely valuable.

The mountaineering abilities of the guanaco, its domestic descendants, and, for that matter, its close wild relative the vicuña, extend even to their metabolic processes. Their blood cells, small compared to those of humans, are numerous, providing a greater capacity to carry oxygen. Moreover, the hemoglobin in their blood has a relatively high affinity for oxygen, further boosting the talent these creatures have for operating under full speed at altitudes where exertion would leave many other animals gasping.

The guanaco, llama, alpaca, and vicuña belong to a family noted for its toughness and ability to get along under the most difficult conditions. The Old World members of the group, the camels, carry out the tradition in the deserts. Both of the two types of domestic camel can negotiate the most parched wastelands, the two-humped or Bactrian camel *(Camelus bactrianus)* where it is cold, the one-humped or dromedary *(C. dromedarius)* in regions where it is searingly hot. The endurance of the camel without water in the desert is legendary; the reasons for it are not fully understood. It seems, however, that the camel lasts so long without drinking because it uses water with nearly ultimate efficiency. Its urine is highly concentrated, but that is not particularly unique for desert animals. The camel, though, has other, rather unusual adaptations for living in the hot waterless places. Its woolly coat insulates it from the sun's heat. It can lower its body temperature during the cool desert night to as low as 90° F. It warms up slowly, and it is not until its body temperature exceeds 104° F. that it must begin to cool down. It can obtain water from its food, especially in the cooler winter when one can go months without drinking. Legends of water stored in the hump or stomach are untrue. For thousands of years—perhaps six thousand in the case of the dromedary—the camel's toughness and ability to withstand heat and lack of water have enabled the people who live with it to survive in regions that otherwise would not be populated by humans.

221 *and* **222** *overleaf. This young alpaca (Lama pacos) has been selectively bred for the long wool that will be sheared from its sides when it matures—a wool of unmatched quality that will be made into coats costing several thousand dollars. Once, garments of alpaca were worn only by Inca royalty. The alpaca and llama are domesticated forms of New World camels; two other kinds, the guanaco and vicuña, survive in a wild state in the South American Andes. The exact lineage of the alpaca and llama—indeed, even whether they should be considered true species—is a mystery that probably will plague science forever. Neither the alpaca nor the llama existed in the wild when the Spanish conquistadores arrived, and archaeological discoveries suggest they had been tamed and bred many centuries before the Inca Empire, probably from the wild guanaco. Moreover, all four New World camels interbreed and produce fertile offspring.*

226-227. *The Altai Mountains rise in the distance as a herd of wild two-humped camels (Camelus bactrianus) crosses the Gobi Desert of Mongolia. There are two species of Old World camels, the one-humped dromedary, which originally came from Arabia, and the two-humped Bactrian camel from Chinese Turkestan and Mongolia. Both have been widely domesticated as beasts of burden, and only the Bactrian camel still survives in a wild population — numbering about 900 animals on both sides of the China-Mongolia border. In summer, the wild camels climb into the mountains as high as 11,000 feet to escape the heat, returning to the grassy steppes and desert in winter. Traveling in groups of perhaps twenty females and*

young, led by a male, they forage in the morning and evening for grass, herbs, branches, and, in autumn, the fallen leaves of poplar trees. Heavy hunting for its meat and hides, and competition from domestic animals for scarce water and pasture, are blamed for the wild camel's decline. But this rare species is now strictly protected by both nations.

228-229. *A herd of Indian wild asses* (Equus hemionus khur) *on the Little Rann of Kutch, a vast salt-impregnated wasteland near the Pakistan frontier. This is one of five subspecies of the Asiatic wild ass; one race, in Syria, is feared extinct, and all the others are considered rare. Competition from domestic livestock for badly overgrazed forage, uncontrolled slaughter for its meat and because its testes were thought to be a powerful aphrodisiac, and its capture for use as a draught animal exterminated the Indian wild ass from most of its historic range. But a thousand wild asses still survive on the Little Rann of Kutch, which lies only a foot or two above the Arabian Sea; there they have been rigorously protected for*

decades. Moreover, local people are
strict veg_____rians.
230 over_____. The only surviving
rac___of truly wild horse is
Przewalski's horse, which was once
widely distributed across Asia.
About two hundred of its kind are
found in zoos around the world, and
it is barely possible that a few still
exist in the Gobi Desert of China
and Mon____lia. The so-called wild
horses that range the western plains
of Nort__America are feral descendants o__cowboys' steeds. And the
Camargue horses that splash across
the shallow lagoons of that marshy
island on the Rhone delta of France
likewise are domestic animals gone
wild_although they are a fairly recent
descendant of a wild horse that was
crossed with Oriental blood.

Photographic credits for the preceding illustrations:

Unique Giants

High upon Kenya's Aberdare Mountains, towering
almost 13,000 feet into the equatorial sky, lie rolling
moorlands, swept by winds that play with the mist
clouds and fray their edges to shreds. The air is cool,
even chill, and the moors are laced with icy streams
glistening under the roof of heaven. Scattered about
the moors are outcrops of rock and, breaking up
the expanses of low vegetation, thickets of giant heath,
taller than a man and bearded with lichens.
The gray-green vegetation of the thickets has a
ravaged look, as if torn and splintered by some
immensely powerful force. That, in fact, is exactly what
has happened, for elephants climb the steep, slick trails
into the highlands and feed in the foggy heath groves.
Unable to jump, the elephants nevertheless are agile
enough to climb the steep slopes that separate the
moorlands from the forest below. Once they gain the
top, they bulldoze into the thickets, white tusks
gleaming amidst the dark foliage. The sight evokes
images of Pleistocene times, some 50,000 years ago,
when pachyderms roamed subglacial landscapes quite
similar to the rolling uplands of the Aberdares.
Largest living mammals outside the sea, the elephants
serve as reminders that, in the Pleistocene, prodigious
size was not uncommon among land mammals. In the
world of nature, size is a signal advantage for a
herbivorous creature; colossal bulk and the strength
that goes with it can render a creature such as the
elephant invulnerable to any land predator. But today
the very size of the elephants and a few other giants
has become a liability, because space is at a premium,

especially where these giants live. With the possible exception of the giraffe (*Giraffa camelopardalis*), the last giants have been pushed by expanding human populations into a few fragile havens in the jungles of Southeast Asia and the forests and grasslands of Africa. After existing unchanged for a million years, the Javan rhinoceros (*Rhinoceros sondaicus*) numbers no more than a score of animals clinging to life in a small reserve in Java and at best a handful roaming the backwaters of Indochina. The Sumatran rhino (*Didermocerus sumatrensis*) at one ton in weight the smallest of the giants, and heir to a line 40 million years old, is almost as rare. Perhaps less than 150 of these shaggy-eared forest rhinos survive, scattered through Southeast Asia. In the swampy grasslands and jungles of the Kaziranga National Park of Assam, India, and in the Royal Chitawan National Park of Nepal, live most of the world's 1200 great Indian rhinos. Like the Javan species, its close cousin, the Indian rhino (*R. unicornis*) is armored with heavy plates of thick skin, joined by thin folds that permit graceful, free movement. Given to frequent squabbles, Indian rhinos fight each other with slashing, and often fatal, swipes of incisors lengthened into razor-sharp tusks. The Indian rhino seldom employs its horn, which, as in other rhinos, is not a true horn with a bony core but is made of matted keratin fibers, like the material that forms a hoof. Perhaps because the Indian rhino prolongs its spectacularly violent mating act for the better part of an hour, rhino horn, from all species, is prized in Chinese folk medicine as an aphrodisiac, and sells for between $1000 and $2000 a pound. This belief is unfortunate, for it makes rhinos a prime target for poachers. In most Oriental cities with large Chinese populations, one finds in medicine shops horn of rhinos from as far off as Africa. Both African rhinos, the black (*Diceros bicornis*) and the white or square-lipped (*Ceratotherium simum*), carry two horns, which may reach a yard or more in length. Weighing more than 6000 pounds, the white rhino is the largest living rhinoceros, but a hornless rhino of about 30 million years ago towered twice as high and may have been the largest mammal ever to walk the earth.

Unlike the others, which feed mainly by browsing, the white rhinoceros is primarily a grazer and crops tough savanna grasses with the aid of the hard-edged upper lip of its squared-off snout. Two thousand miles divide the last two populations of white rhinos on earth, for of the few thousand left, some live on protected South African reserves, the others under questionable protection on the grasslands of Uganda, the northern Congo, and southern Sudan.

The windswept thorn scrub of Kenya and Tanzania is the heart of black rhino country, but this myopic beast inhabits several other regions scattered about sub-Saharan Africa. Tough as the whistling thorns among which it lives, the black rhino has a reputation for truculence, but this is partly the fault of its miserable eyesight, and often what appear to be thundering, snorting charges merely represent an attempt by the creature to take a close-up look. Surprised in the high grass, a black rhino like as not will face in the direction of an interloper, its piggy eyes squinting, the twitching of its nostrils wrinkling its snout above its prehensile upper lip, and ears flared. After a few moments, the rhino may wheel about and trot off. Then again, it may explode in a headlong charge.

More than 10,000 black rhinos live in Africa, but they depend almost entirely on sanctuaries, which they often share with the colossus among the giants, the African elephant *(Loxodonta africana)* of the bush country. A big bush elephant can reach six tons in weight and thirteen feet at the shoulder, twice the size of some members of the same species living in the green confines of the forest. When the need arises, the elephant can move its mountainous gray bulk over the ground at 25 miles an hour.

A slow-motion study in violence, the advance of an elephant herd eating its way through a grove of acacias in the African bush is unhurried but inexorable, and guaranteed to make a watcher feel puny, so casually do the great beasts commit acts of Herculean destruction. The elephants' great trunks, whose two-fingered tips are sensitive enough to pick up a peanut, gently wrap around sturdy tree limbs, then tear them away more easily than a man could pluck a rose. If leaves are too high, the trees that bear them are nudged by broad gray foreheads until wood splinters and they crash to the ground. An African elephant can consume up to 600 pounds of vegetation daily, so not surprisingly, when confined to parks, elephants literally eat themselves out of a home. Their destructive feeding has become an increasingly critical problem in Africa. Elongated upper incisors that grow throughout life, the tusks can reach a length of a dozen feet in African elephants. Although nominally outlawed, killing of elephants for their tusks goes on in most of their African range, with the ivory going to markets in Hong Kong and mainland China. Of itself, the poaching might not imperil the existence of the species, but in the face of declining habitat, the African elephant cannot long endure the poisoned arrows and traps of poachers. The slightly smaller Asiatic elephant *(Elephas maximus)* also suffers severely from the constriction of its habitat, but in Asia, where wild elephants have been trained as beasts of burden for 5000 years, they seem more adaptable to life surrounded by people. Asiatic elephants, moreover, are not nearly as persecuted by ivory hunters, because they have much smaller tusks. The hippopotamus *(Hippopotamus amphibius)* of Africa also carries formidable tusks, which are not incisors but canines in the lower jaw. When angered, the hippo open its jaws in a cavernous yawn, baring its huge weapons, which it can wield with lethal effectiveness. Most of the time, however, hippos are the picture of luxuriant loafing, particularly during the day. Bunched together in the mud like hogs, they bask on sandbanks, their dark hides tinged red by protective mucus, secreted by glands in the skin. Or else they float just breaking the surface of quiet water, an activity for which their eyes, ears, and nostrils are suitably positioned atop the snout and head.

Although the Indian rhino excels as a swimmer, and elephants revel in bathing, none of the giants is as aquatic as the hippo, which even is born underwater. Able to walk on the bottom, the hippo can stay under for up to six minutes.

After dark, however, the hippo is transformed. It heaves its body from the water and, leaving its platter-sized, four-toed prints in the mud, briskly heads overland. Along age-old trails, hippos may travel miles to their feeding plots. In the course of a night, 100 pounds of grass and other fodder can disappear down the maw of a single large hippo. If farmers have planted crops on hippo feeding grounds, the devastation is swift, and so is the reprisal. The last terrestrial giants have no way of knowing that they evolved in a world when humans were scattered and few, and no plow had broken the earth. They cannot know that the days when they were lords of the land are over, and that their fate is now entirely in the hands of a pygmy called man.

237. *When a bull hippopotamus* (Hippopotamus amphibius) *opens its cavernous mouth in a "yawn," exposing sharp incisors and huge lower canine tusks that may be twenty-five inches long, it is not a sign of laziness. By showing off the formidable weapons that jut from its powerful jaws, the hippo is trying to intimidate a rival. But such threat displays as often as not incite rather than prevent fights, and a battle between two hippos can be awesome and bloody. For an hour or two the hippos rush at each other with gaping mouths, sending waves rushing across the lake or river as they attempt to drive those great canine teeth—once used to make human dentures—through the thick hide and into the heart of their opponent.*
238 *overleaf. A single horn and thickly folded, tubercle-covered skin that gives it an armored appearance identify the great Indian rhinoceros* (Rhinoceros unicornis). *Only a few hundred individuals of this endangered giant still survive in sanctuaries in India and Nepal. Weighing two and a half tons, the great Indian rhinoceros is the second largest of the five species of rhino found in Africa and Asia. The female bears a single calf after a gestation period of sixteen months, and the newborn rhino weighs 140 pounds and has all the skin folds and rivetlike protuberances of an adult, lacking only the nose horn. Over its first year of life, drinking twenty-five quarts of milk a day from its mother, the calf will multiply its weight tenfold at the rate of six pounds a day. It will not be weaned until it is two and a half years old.*

242 overleaf. *With its hind feet reaching ahead of the front feet and its long neck swinging like a pendulum, the giraffe* (Giraffa camelopardalis) *can lope tirelessly across the East African plains at speeds of up to 30 miles an hour. Running is the giraffe's primary defense against its only natural enemy, the lion, but if cornered it can crush the big cat's skull with blows from its front legs.*

240-241. *Worrying the desiccated remains of one of its own, ripping apart an acacia tree to obtain part of the 300 pounds of food it requires every day, bathing with dust—all are routines in the daily life of the African elephant* (Loxodonta africana), *the largest land mammal on earth. In particular, elephants spend a great deal of time washing, powdering, and massaging their skin, which is unusually sensitive for a beast that weighs six tons.*

Photographic credits for the preceding illustrations:

Life on the Peaks

Although many mammals roam the world's high places, one group of horned, hoofed creatures in particular has made the windswept, stony heights its special playground. Native in one form or another to Eurasia, northern Africa, and North America, the goats, sheep, and a few related beasts clatter over all but the very highest reaches of the mountains with a wild freedom that makes them seem kin to the wind. Superbly equipped for life atop the crags, these mountaineering mammals feed, breed, and even rear their young on landscapes of incredible harshness. The four-chambered stomachs of these ruminants can glean enough nourishment from rubbery scrub, wisps of grass, and scraps of lichen to thrive in the marginal environment of the peaks, where other large mammals would starve. For example, the mouflon *(Ovis musium)*, wild sheep of the Mediterranean, can subsist even on the plant known as deadly nightshade.

Amazingly sure of foot, these creatures of the alpine solitudes have a dizzying aptitude for cavorting on eminences above the clouds. Persian wild goats *(Capra hircus)* leap about the boulders almost 14,000 feet up in the barren mountains west of the Caspian Sea, where Iran, the Soviet Union, and Turkey meet. The big Siberian ibex *(C. ibex sibirica)* goes even higher and, like others of its species, easily perches on any pinnacle with enough room for its four feet. Bighorn sheep *(Ovis canadensis)* survey the Rocky Mountains from lookouts 10,000 feet high. In the Himalayas, blue sheep *(Pseudois nayaur)*—not exactly true sheep and not blue, either, but gray—loll on grassy slopes at 18,000 feet.

Of them all, however, the true king of the mountain is the North American mountain goat *(Oreamnos americanus)*, a bearded will-o'-the-wisp as white as the snow that swirls about its native western mountains in the winter. Once sighted in the heights, the goats are given to vanishing with ghostly abruptness, then suddenly reappearing at the edge of even higher peaks. The mountain goat is not a true goat, but along with its cousins the Eurasian chamois, and the serows, gorals, and takins of Asia, is classified as a goat antelope, because it possesses traits of both. Scientists have trouble classifying some of the mountaineering mammals. The blue sheep, for instance, is not quite a sheep, and not a goat, but a little like both, so in the end its name is based upon its general appearance.

The backgrounds against which the mountain goat is pictured tell much about the brutal conditions with which it must cope. Unlike the bighorns, which descend to the shelter of lower slopes in the autumn, and the corkscrew-horned markhor goats *(Capra falconieri)* of Asia, which emerge from the heights to feed on live oak when the snow flies, the mountain goats seldom stray below the timberline. There are exceptions, of course, but mountain goats generally head for the trees only in the spring, when the sweet, green shoots lure them from their high havens. The goats negotiate the steep slopes and sheer, gray headwalls of their world with stiff-legged deliberation, pausing frequently to consider the next move. Traveling this way a goat can ascend nearly vertical palisades, when necessary pulling its 300-pound body from ledge to ledge with its black forefeet. If a goat reaches a dead-end ledge, it rears upon its hind legs, whirls about, and returns the way it came, often by stupendous leaps from rock to rock. The fancy footwork of the goat is due in large measure to the structure of its hooves; it and the other members of the group have two on each foot. In the center of each hoof is a spongy, elastic pad, like a tire tread, which provides traction, while the hard-edged rim of the hoof catches in minute clefts and crevices—an arrangement shared by other mountain mammals.

The chamois *(Rupicapra rupicapra)*, which roves mountain ranges from the Pyrenees to the Caucasus, and has been transplanted to New Zealand, is more graceful than its American relative. But it has the same uncanny ability to materialize seemingly out of nowhere. One moment a talus slope appears empty, the next a chamois is there, scanning the slopes below.

If it senses danger, however, the chamois moves with dazzling speed, leaping and bounding over chasms so lightly it seems to fly. If cornered, a chamois fights with vicious thrusts of its foot-long horns, but rarely are these weapons used so murderously as in rutting battles between the males.

Chamois, and mountain goats, sometimes fight to the death in such combats, and once the weaker duelist gives ground, he may be pursued ferociously, even knocked spinning from a cliff. Male mountain goats, moreover, do not reserve their aggressiveness for other adult males, but are notorious kid killers, perhaps the worst enemies of their own young.

The rutting combats of most other animals in this group, on the other hand, are ritualized to prevent serious injury and death. The crash of bighorns hammering against each other's huge, curled horns sometimes can be heard a mile away—but the blows are always horn on horn. The horns also armor their owners against even pile-driving blows, which also are cushioned by the skull. Bighorns have a tendency, like an experienced boxer, to "roll with the punch." Males with horns markedly different in size seldom fight.

The males of most of the true sheep and goats carry horns that are massive in relation to their body size. The Nubian ibex, found in Israel and the Sinai, as well as Africa, has horns almost four feet long. Yet this smallest of ibexes is not much more than knee-high to a man at its shoulder. The Persian wild goat, which weighs less than 100 pounds, has saber-shaped horns that sometimes reach a length of more than five feet. And even a mouflon, little more than two feet high at the shoulder, can have horns a yard long.

The most impressive horns of all, however, belong to the argali sheep *(Ovis ammon)*, an awesome creature four feet high at the shoulder, larger than a bighorn. Some argalis carry horns that, flaring outward in a wide spiral, approach six feet from tip to tip. The argalis inhabit some of the bleakest mountains in the world, in the cold interior of central Asia, and in their isolation are symbolic of all the mountain animals, for they are creatures of a world that, until recently, existed on a plane above humanity. As long as the mountaineering mammals remain in the wild, they are assurance that somewhere on this earth, a living creature remains in splendid isolation.

249. *A Dall sheep ram* (Ovis dalli) *snorts a warning. Wild mountain sheep—rams in particular—are customarily silent animals, but on their summer pasture the ewes and lambs in a herd carry on a noisy conversation that suggests a flock of domestic sheep. This species is named for the American naturalist William Healey Dall, who was a student of Louis Agassiz and one of the first scientists to explore Alaska at the time of its purchase from Russia by the United States.*

250 *overleaf. The white coat of the Dall sheep harmonizes with its chosen home—the snowy mountain heights of Alaska and northwestern Canada. This is a summer bachelors' club of rams of all ages, for they play no role in the raising of lambs. But such peaceful coexistence will end in late autumn, when jealous competition for the attention of ewes is renewed. The slender, flaring horns of Dall sheep contrast with the massive, tightly curled horns of the bighorn sheep that claim the Rocky Mountains to the south.*

252 *second overleaf. A sifting of hard snow crystals salts the face of a 300-pound Rocky Mountain bighorn ram* (Ovis canadensis) *during a subzero snowstorm. The chips in his huge horns are the scars of head-butting battles, and the horn tips are splintered from violent clashes with other big rams. The deep creases were formed each autumn when the horns temporarily stopped growing, and the segments between the creases represent the twelve years of his life. The annual growth gets smaller as the ram grows older, and this monarch is nearing retirement, for few bighorns live beyond the age of fourteen.*

254 *third overleaf. The master of a harem of bighorn ewes routs a young rival. Negotiating such precipitous cliffs is not a great challenge for a mountain sheep. Its hoof has a hard outer edge and toe that grips in loose dirt or rock cracks, plus a resilient pad at the back that provides traction on smooth surfaces. Moreover, to the ram's eye there usually are well-defined, if zigzag, paths across such jagged rock faces. And if not, the bighorn can plunge down a rock chute, hurtling from niche to niche in a controlled fall with the ease that comes from superb traction and balance.*

256 *fourth overleaf. Fights between two male ibex* (Capra ibex) *are spectacular affairs, but they are usually only ritualized play between two peers rather than serious combat for domination of a harem. Rising on their hind legs, these bearded wild goats will clatter their four-foot-long, saberlike horns together; or they will wrestle with horns hooked, pushing with their foreheads. Ibex dwell high in the mountains of Eurasia and northern Africa, browsing alpine meadows and resting in the shade of rock overhangs. They were nearly exterminated throughout much of their range because superstitious people—as late as the nineteenth century—believed that their horns, blood, heart muscles—even their feces—could cure countless ailments. But the ibex has been successfully reintroduced in many parts of the Alps.*

258 *fifth overleaf. The highest crags of North America's highest mountains are where the mountain goat* (Oreamnos americanus) *is likely to be found in summer. There, far above the haunts of traditional enemies—wolves, grizzly bears, cougars—the nanny and her one or two kids are safe from all enemies but soaring golden eagles. And it is a rare event when one of the huge raptors succeeds in carrying off a kid. As winter approaches, and the mountain goat descends to forested valleys to escape deep snows, it acquires a heavy coat six to eight inches long on top of a four-inch layer of woolly underfur.*

Photographic credits for the preceding illustrations:

Mammals in the Sea

Watching a great whale rise from the depths of a calm
sea is like witnessing the birth of an island. One
moment the sea is smooth, stirred only by light swells
that slide gently across its surface, or perhaps by the
occasional splash of a seabird knifing into the water
after a fish. Then the ocean parts. A dark and mountainous form, encrusted with a white mottling of
barnacles, heaves above the surface. Water drains in
torrents from the crest of its back and sluices into the
foam boiling around it. Overhead, like clouds of steam
above a volcano, hangs the condensation from the sea
beast's exhalations, which fill the air with a mighty
rushing sound. It is an experience to make one feel
fragile and insignificant, and to prompt pondering about
man's place on earth.

The great whales are the most spectacular of a group of
mammals that have adapted to the world of water. This
assemblage also includes the smaller whales, dolphins,
and porpoises, the seals and their kin, and the dugongs
and manatees. In a sense, they have come full circle,
for they have returned to the environment from which
their remote cold-blooded ancestors crept and, indeed,
from which all life arose. Most of them are marine, but
a few, such as the weak-eyed river dolphins (Platanistidae) of Asia and South America, and Siberia's Lake
Baikal seal *(Pusa sibirica)*, spend their lives in fresh
water. Although the mammals here belong to three
unrelated orders, they share a parallel history. For
reasons only to be surmised, the ancestors forsook the
land and committed themselves to life in the water—
probably because the pressures of competition or the

need for a new source of food forced them to.
The whales, dolphins, and porpoises—together called
cetaceans—made the transition in the Eocene Period,
at the dawn of the Age of Mammals more than 45 million
years ago. The dugongs and manatees, comprising the
order of sirenians, accomplished it at about the same
time. Not for another 20 million years, in the Miocene
Period, did the seals and their relatives appear in the
sea. They are sometimes grouped together in a single
order, the pinnipeds—meaning "wing-footed"—but all
of them may not have evolved from the same ancestors.
The true seals, streamlined for swimming to the degree
that they have lost external ears, possibly hark back to
an otter-like ancestor. The fur seals, sea lions, and
walruses, on the other hand, may descend from creatures close to bears.

The sirenians probably arose from the same rootstock
as the elephants, but the origin of the whales and their
cousins is elusive. Students of evolution have seen in
the cetaceans signs pointing back both to carnivorous
ancestors and herbivores. Some of the earliest whales
had teeth similar to archaic meat eaters. Yet chemical
analysis shows that modern cetaceans contain some
proteins similar to those in sheep, cattle, and camels.
No matter what ancestors produced the cetaceans and
the other animals considered here, any resemblances
they have to distant kin on land are only fleeting echoes
of ancient relationships. The pinnipeds, for instance,
may be offshoots of the carnivores, but living in the
water has sufficiently reshaped the seals and the rest
of their flippered band to confer upon them a distinct
identity. The ways in which such creatures have adapted
to life beyond the world's shorelines merit profound
consideration, if not awe, because they all have managed
to cope with an environment almost as hostile to a
mammal as the cold blackness of space.

As a matter of course, these animals go for long
periods without breathing the air upon which they,
like all other mammals, depend. The sirenians sometimes stay below for a quarter of an hour as they pull
eelgrass, water hyacinths, and other aquatic plants
into their small mouths with the aid of a fleshy and
flexible upper lip. The sea otter *(Enhydra lutris)* scours
the bottom down to one hundred feet, probing with
stubby forepaws for the shellfish and sea urchins it
relishes. A month-old harbor seal *(Phoca vitulina)* can
dive for up to four minutes on a single breath. A year-

ling of the same species can hold its breath for twenty minutes. The southern elephant seal *(Mirounga leonina)*, a giant twenty feet long weighing up to 8000 pounds, need not come up for a half hour after it dives. The Weddell seal *(Leptonychotes weddelli)*, a 900-pound beast of the south polar regions, spends more than half the year in the sea under several feet of ice, and sometimes must hold its breath for an hour as it travels miles between breathing holes. This seal customarily cruises at depths approaching 1000 feet, but it may dive to twice that depth. The legendary sperm whale *(Physeter catodon)* routinely searches the dim reaches 1500 feet down for squid and not rarely descends to almost 3000 feet. Several years ago, one of these behemoths was found entangled in a transoceanic cable at 3200 feet. When necessary, as in a titanic battle with a huge squid, a sperm whale can remain under for an hour and a half. Whales may stay down longer and range even deeper than we suspect, for although they have been pursued over the surface by humans for centuries, we really know very little about how they behave beneath the waves.

To remain under very long without breathing, a sea mammal must use oxygen with exceptional economy, and in fact this ability is one of the traits unifying the orders which have left the land behind. The sirenians and whales promote oxygen conservation with their relatively low metabolism. In terms of calories per kilogram of body weight, the basic metabolic rate of a whale is one-fifteenth that of a human, and the blood of a typical whale holds sufficient oxygen to keep its bodily processes going for more than an hour. Moreover, the deep-diving sea mammals have a generous supply of blood for their size. A seal, for example, has one and a half times as much as a land mammal of equal bulk. Sea mammals also have an abundance of myoglobin, the oxygen-carrying protein that serves the same function in the muscles as hemoglobin in the blood. It all means that their bodies store vast amounts of oxygen.

Even more telling is the astonishing efficiency with which sea mammals operate during a dive. They reduce, even shut off, all bodily activities not essential beneath the surface of the water. By selected constriction of the blood vessels, the flow to the kidneys and digestive system, for example, is slowed drastically, while blood continues to rush unimpeded to the brain, spinal cord, and heart muscles. Thus the oxygen in the blood is reserved for the organs which must function during the dive. At the same time, the heartbeat decreases; in the bottle-nose dolphin *(Tursiops truncatus),* for instance, it is reduced to as much as half the normal rate. When the gray seal *(Halichoerus grypus)* dives, its heart beats 10 times a minute, as compared with 100 to 150 times a minute at the surface. Beyond this, the sea mammals seem rather insensitive to the buildup of carbon dioxide, which in land mammals quickly triggers impulses in the respiratory center of the brain that result in the act of taking breath.

The ability to clamp the nostrils tightly shut underwater, closing off the respiratory system, is an enormous advantage for a mammal in the water. For this purpose, the nostrils of pinnipeds are narrow slits. Those of the sirenians and cetaceans are round orifices ringed by a muscular rim. During the course of evolution, some major rearranging was also accomplished in the nasal tract of the cetaceans. The nostrils have been moved from the snout to the apex of the head, forming a single blowhole in the toothed whales, paired in the baleen whales. As a result, a whale can breathe with only the tip of its head above the surface. Underwater, a whale opens its mouth without danger of drowning, because the nasal and throat passages are separated, so that the lungs open to the outside only through the blowhole.

Many other changes, some quite obvious, have occurred in those mammals that truly have made the water their home. Most recognizable is that they have taken on the so-called fusiform body shape typical of fish, at its sleekest, perhaps, in creatures such as the killer whale *(Orcinus orca).* Even such bulky and ponderous animals as the walrus *(Odobenus rosmarus)* and dugong *(Dugong dugon)* streamline into a torpedo shape as they move through the water. The blubber that insulates them smooths out, reshaping their lumpy outlines into symmetrical contours. The density of water has had other influences on the sea mammals. The support provided by water has allowed some of them to reach enormous size. Only the aquatic environment could produce an animal with a body more than 100 feet long, weighing 170 tons. That creature is the blue whale *(Balaenoptera musculus),* the largest animal that has ever lived. Ironically, the blue whale may vanish because of its size, which, marked this mighty sea mammal as a prime target of whalers—men as rapacious as any of the beasts under the sea.

265. *Using its powerful forelegs for propulsion, a Galápagos fur seal* (Arctocephalus australis galapagoensis) *plunges into a grotto on James Bay, on San Salvador Island. The fur seals that bred in immense numbers on islands around the Southern Hemisphere were nearly annihilated in the nineteenth century, slaughtered for their treasured pelts. In the first two years after their discovery, for example, 600,000 sealskins went to market from the South Shetland Islands. In 1821, a single ship arriving in London from Antarctic sealing grounds carried 400,000 pelts. What saved most of the several species of southern fur seals was economics: two few remained to make further harvests profitable. In the case of the Galápagos fur seal, the fact that a few individuals retreated into lava caves in daytime prevented its extinction.*
266 *overleaf. This bull Steller's sea lion* (Eumetopias jubata), *surrounded by his harem, is ten feet long and weighs 2000 pounds. Grown fat on a diet of fish, this inhabitant of Pacific beaches from the Aleutian Islands of Alaska to Japan and California is the largest of several species of sea lions. It is named for Georg Wilhelm Steller, the German who was the first naturalist to reach Alaska in the eighteenth century.*
268 *second overleaf. At sunset on Hood Island, a Galápagos sea lion bull and cow* (Zalophus californianus wollebaeki) *luxuriate in the falling spray from a blowhole. Like fur seals, sea lions belong to the family of eared seals. They can stand on four "legs" and almost gallop on land.*

270 *top left. Sea lions soak up the equatorial sun burning down on the Galápagos lava. Unlike fur seals, sea lions were not hunted primarily for their pelts, since their coat has no underfur. Instead, they were slaughtered for food and rendered for oil and their skins were processed into low-grade leather.*

270 *above left. The aptly named leopard seal (Hydrurga leptonyx) patrols Antarctic ice floes, waiting for Adélie penguins to enter the water. Its prey is quickly dispatched, shaken out of its skin, and swallowed whole. So fierce that they are avoided even by packs of killer whales, leopard seals are longer than all other seals except the elephant seal. Powerful swimmers, they are able to shoot out of the water and land on ice eight feet above.*

270 *top right. Harp seals (Phoca groenlandicus) are true seals that move on land only with considerable difficulty. Their annual oceanic migrations cover 6000 miles, and scientists calculate that a harp seal must consume a ton and a half of fish and crustaceans to fuel its journey. Harp seals bear their young on pack ice off the coasts of Labrador, Newfoundland, and Greenland, and the slaughter of newborn pups for their pure-white coats is a continuing source of international controversy.*

270 *above right. Soon after the birth of her single pup, which weighs less than five pounds, a sea lion cow will mate again. The pups, which bleat like lambs, spend their first three months nursing and frolicking in*

tidal pools. The mortality rate is
high: pups often fall into the sea,
are unable to scramble back onto the
rocks, and drown; others are crushed
by bulls fighting over a harem of
ten to twenty cows.
271. Skilled swimmers, sea lions
can dive to depths of more than 300
feet and remain submerged for
fifteen minutes. But deadly enemies
are waiting offshore: sharks and
killer whales.

272 overleaf. Young bull elephant
seals (Mirounga leonina) gather in a
knotlike formation on a beach in the
South Shetland Islands, between
Cape Horn and Antarctica. Easier to
kill than a whale, these blubbery
monsters—a mature bull may carry
four tons of weight on its fifteen-
foot hulk—once were butchered by
the tens of thousands and boiled
down for their oil.

274. *With surprising agility, a bull elephant seal has dragged itself to the luxury of its mud wallow on Campbell Island, south of New Zealand. On the breeding grounds, old bulls have a busy time keeping possession of the much smaller and faster cows in the harem, for the opportunistic younger bulls in the vicinity are always awaiting a chance to mate with unguarded cows.*
275. *The northern elephant seal* (Mirounga angustirostris) *of California waters nearly passed into extinction in the 1890s. By the time it was given complete protection from hunting, fewer than 100 animals survived. Today there are 15,000 on the Channel Islands off Los Angeles. The northern species has a longer "trunk" than its Antarctic cousin—*

that snout that dangles limp most of the year but is inflated during the excitement of the mating season. Elephant seals haul out on land twice a year—once to breed, and again a few weeks later to molt. While ashore they fast, living off their great fat reserves built up by feasting on cuttlefish and squid caught in the ocean depths.

276. *The ivory tusks of a walrus (Odobenus rosmarus) grow throughout the animal's life, and those of a mature male can reach a length of forty inches and weigh twelve pounds. The walrus uses its tusks to haul its two-ton hulk out of the sea and as a weapon—and possibly to dig out mollusks from the floor of the ocean.*

277. *Their skin baked red by the Arctic sun, walruses sprawl across a rocky island in the Bering Strait separating Alaska from Siberia. The hide of a walrus may be an inch thick, and is so tough it was used in the construction of the Cologne cathedral in Germany in the thirteenth century.*

278 *overleaf. Two walrus calves nuzzle each other, showing the stiff bristles that, when they begin to fend for themselves after nursing for two years, will be an important tool in feeding. A walrus has a mustache of some four hundred sensitive bristles that are used not only to rake up food on the muddy bottom but also to hold a mollusk shell while the animal is sucked out. A young calf will cling to the back of its mother with its front flippers as she dives after food, and in this position it may be vulnerable to an attack by killer whales. But walruses have few enemies other than man; polar bears sometimes kill calves on land or ice, but are no match for protective mothers in water.*

281. *Floating on its back in California's Monterey Bay, a sea otter* (Enhydra lutris) *is impervious to the sharp spines of the sea urchin on which it is feeding. Sea otters are a rarity among mammals: they use tools! To smash open the hard shells of mollusks, the otter lays a stone on its chest and uses it like an anvil. The stone is then carried under the otter's arm as it dives as deep as 130 feet in search of another clam, snail, or urchin. Because it inhabits cold Pacific waters, the sea otter requires a prodigious amount of food—up to twenty pounds a day— to maintain its body temperature.*
282 *overleaf. A common dolphin* (Delphinus delphis) *races alongside a yacht in the Galápagos Islands. Eight feet long, capable of attaining a speed of 30 miles an hour, this is truly the common dolphin of temperate seas around the world; herds of* Delphinus *numbering in the thousands can literally churn the ocean to froth as they pursue squid, baitfish, and flying fish.*

280. *A 1500-pound northern manatee* (Trichechus manatus) *can remain submerged for sixteen minutes as it feasts on aquatic vegetation. It pushes plants toward its mouth with its flippers, picks them up with a split lip that is used like a forceps, and works the food into its mouth with the stiff bristles on its muzzle. In cold weather, Florida manatees congregate around warmer springs in rivers and even city water outlets in Miami.*

284. *In the clear waters off the Hawaiian island of Maui, a calf humpback whale* (Megaptera novae-angliae) *stays close by the side of its mother. Some two hundred of these rare whales, famous for their "singing," use the warm near-shore waters of Hawaii as a mating ground and as a nursery, remaining there from February to June, when they depart for nutrient-rich feeding grounds in cold northern seas.*

285. *Each wartlike knob on the head of this humpback calf contains a single sensitive vibrissa that is similar to a cat's whisker. The baby whale, sixteen feet long at birth, is not yet parasitized by the barnacles that bedeck adult humpbacks —as much as a half ton of them affixed to a single whale.*

286 *overleaf. In a burst of spray, a right whale* (Balaena glacialis) *explodes from a bay off Punta Váldez on the coast of Argentina, hurling itself into the air with such momentum that it somersaults. A whale will breach repeatedly, a score or more times, always landing on its back or side with a thunderous whack. Just why whales breach is not understood; it may be done to dislodge parasites—or just for sport.*

Notes on Photographers

Durward L. Allen (135), veteran wildlife photographer, is Professor of Wildlife Ecology at Purdue University.

Charles Andre (91) is an environmental consultant for Brighton Engineering in Frankfort, Kentucky. His photographs have appeared in Sierra Club publications.

Fred Baldwin (170) lives in Georgia and specializes in photographing the southeastern United States and the Arctic.

Jen and Des Bartlett (286) are Australians known for wildlife films (including the notable *Flight of the Snow Goose*) and for stills made on every continent.

Erwin A. Bauer (154, 214) has worked as a freelance photographer in northern California since 1948. He is the author of *Living Water* and *Tideline*.

Wolfgang Bayer (95, 119, 154, 242) is a wildlife film producer who has made television films for the National Geographic Society, Walt Disney Productions, and *Wild World of Animals*.

Rajesh Bedi (146) is a New Delhi photographer whose photographs of Indian wildlife have appeared in international publications.

René Pierre Bille (256) is a Swiss naturalist and film maker who is especially interested in wild animals of the Alps.

Tom Brakefield (120) is a full-time nature writer-photographer. His articles have appeared in numerous periodicals.

Kay Breeden (58) is an Australian wildlife photographer whose work has appeared in *National Geographic*, *Audubon*, and in Chanticleer Press books.

Stanley Breeden (40, 58, 129, 145, 183) has photographed wildlife throughout the world. His pictures have been published in numerous international publications.

Fred Bruemmer (166, 270, 277) was a newspaper photographer and reporter in Canada before becoming a freelancer specializing in the Arctic. He has written 300 articles and five books, including *The Arctic*.

John Burnley (189) lives on Long Island, New York, and has photographed in the Arctic and Alaska.

Fred and Dora Burris (106-107) are writer-photographers whose work has appeared in many national magazines, including *Audubon*.

Ron Church (268, 280) was cinematographer for the original Jacques Cousteau television series. He was also pilot of the Westinghouse Deepstar three-man research submarine.

John Cooke (121) has designed and built various types of photographic equipment. He is an experienced mountaineer and scuba diver.

Jack Couffer (110) is a cinematographer whose credits include Walt Disney's True Life Series and *Jonathan Livingston Seagull*, for which he was nominated for an Academy Award.

Bill Curtsinger (270) is a contract photographer for the National Geographic Society, specializing in marine mammal photography.

Thase Daniel (36, 37, 179), a native of Arkansas, has photographed wildlife in remote parts of many countries. Her photographs have appeared in numerous books and magazines.

Edward R. Degginger (149, 266) is both a professional chemist and a photographer of wildlife. Over 3600 of his pictures have appeared in books, magazines and encyclopedias.

Irven DeVore (39, 47), Professor of Anthropology at Harvard University, has for eighteen years conducted field research on the free-ranging baboons of Kenya.

Nicholas DeVore III (282) freelances for the National Geographic Society. Born in Paris, he now lives near Aspen, Colorado.

John Ebeling (109, 134) is a Minnesota-based nature photographer who is much concerned with protecting porcupines. His pictures have been published in *National Wildlife*.

Harry Engels (120, 254) photographs wildlife of the northern Rockies. His articles and pictures have appeared in many publications.

Francisco Erize (120-121, 206) lives in Argentina and has photographed wildlife in South American jungles, the Galápagos and Antarctica.

Bob Evans (275) is a nature and underwater photographer living in Santa Barbara, California. He heads his own production company, La Mer Bleu.

Jon Farrar (94) has contributed to *Audubon, Natural History* and *Outdoor Life.* His special interest is plants and animals of the North American grasslands.

Jean-Paul Ferrero (36, 49, 63, 64, 66, 117, 168, 182) is a French wildlife photographer whose work has been published in numerous European books and periodicals.

Kenneth Fink (105) became interested in photographing mammals while working for the U.S. National Park Service. He teaches public school in San Diego, California.

Jeffrey Foott (91, 281) is a biologist studying marine mammal behavior. His pictures of wildlife have appeared in *Audubon, National Geographic,* and in Chanticleer Press books.

Michael Freeman (32) lives in London. Much of his photography is done in tropical South America.

Warren Garst (122) is chief cinematographer for the *Wild Kingdom* television series. He has also made films for Walt Disney.

John Gerard (118) is an Illinois photographer who operates one of the largest photographic studio services in the field of wildlife.

David R. Gray (97) is a research scientist with the Museum of Natural Sciences in Ottawa, specializing in studies of mammals in the Canadian High Arctic.

Gary Griffen (184) photographs scenics and animals of the Hudson Valley and Catskill Mountains. He is currently shooting a solar energy film.

Antje Gunnar (258), a mountaineer and world traveler, has had photographs in *International Wildlife* and in National Geographic Books.

Tetsuo Gyoda (44, 45), a widely published Japanese wildlife photographer, is author of *Secrets of Japanese Migratory Birds* and *Animal Families.*

Robert Harrington (237) is a photographer for the Michigan Department of Natural Resources. His pictures have appeared in numerous publications, including *National Geographic.*

George Holton (4, 25, 26, 48, 50, 169, 222, 226, 270, 272, 274, 276) is a New York based photographer whose work has appeared in publications of the Audubon Society, the National Geographic Society and Time-Life Books. His travels have taken him to nearly every part of the world.

Maurice Hornocker (149) began photographing animals while he was a student at the University of Montana. His photographs have appeared in several national publications.

David Horr (46), Professor of Anthropology at Brandeis University, is well-known for his field study of orangutans.

James Hudnall (284, 285) is a cetologist and underwater cinematographer. He has spent three winters documenting the behavior of Hawaiian humpback whales.

M. Philip Kahl's (6, 155, 156, 202) work has been reproduced in many magazines and books. Grants from the National Geographic Society have aided his study of storks and flamingos of the world.

James A. Kern (36) has participated in expeditions to Asia, Central America and the Arctic. He is especially interested in the wildlife of Florida.

Russ Kinne (10, 28, 203) is a veteran freelance natural history photographer and the author of *The Complete Book of Nature Photography.*

Dean Krakel II (186) lives in Oklahoma City and does wildlife portraiture and general photojournalism.

Stephen J. Krasemann's (165, 250) photographs have appeared in numerous wildlife periodicals, books and calendars.

Nina Leen (73, 74, 76-77, 78, 80, 81) was a *Life* staff photographer. She is author of *And Then There Were None: America's Vanishing Wildlife* and *The World of Bats*.

William and Marcia Levy (130) are a husband and wife team from New York. They do films for such television programs as the Time-Life *Wide, Wide World of Animals*.

Sven Lindblad (240) was born in Sweden but now lives in the United States. He has photographed throughout the world, particularly in East Africa, the Arctic and Antarctica.

James Malcolm (131), a graduate student in biology at Harvard University, is engaged in a long-range study of the social behavior of wild dogs in Tanzania.

Tom McHugh (34, 38, 57, 148) is a professional cinematographer who did his first work for Walt Disney Productions. In recent years, he has returned to still photography.

Loren McIntyre (148, 152, 221, 224) is a writer, photographer and film maker. His subjects are South American, with emphasis on the Andes and the Amazon.

Wendell D. Metzen's (96) photographs of wildlife of the southeastern United States have appeared in books, magazines and calendars.

Gary Milburn (249) specializes in photographing the wildlife of South America. He works for the Environmental Protection Agency.

Michael Morcombe (60, 62, 91), naturalist, photographer and writer, has developed techniques for photographing rarely seen Australian bush life. His books include *Wild Australia* and *Birds of Australia*.

James K. Morgan (252) is a wildlife biologist who spent eight years studying and writing about bighorn sheep. His photographs have appeared in *Audubon* and *National Geographic*.

Thomas Nebbia (158, 208, 212) freelances for *National Geographic*, mainly in Germany. Born in Rochester, New York, he now lives in Copenhagen.

Alan Nelson (95) has had photographs published in *Audubon* and in Time-Life publications. He is a biology teacher in Great Falls, Montana.

Kenneth B. Newman (204) is an English painter and photographer who has settled in Africa. He is especially interested in birds and animal life.

Mark Newman (197) is a physician in Sundance, Wyoming. He is a self-taught photographer, specializing in wild animals.

Charlie Ott (91, 188) is on the staff of Mt. McKinley National Park. His pictures have appeared in *Wild Alaska* and many other nature publications.

Richard Parker (30) lives in Missouri and specializes in plant and insect photography.

Rolf Peterson (132) has been studying wolves since 1970. His photographs of wolves and related species have appeared in many magazines including *National Geographic*.

Francis Petter (149) is a French biologist-researcher known for his scientific approach to animal photography.

T. W. Ransom (42-43) began photographing mammals while studying baboons in Tanzania in the late 1960s. His photographs have been reproduced in numerous national publications.

John Reader (240) has co-authored *Pyramids of Life*, a book that examines African wildlife in its ecological perspective. His work has been published in major international magazines.

Michael and Barbara Reed (270) specialize in anthropological photography. Their work has taken them to Central America, Africa, the Galápagos and the West Indies.

Hans Reinhard (108, 120-121) lives on a small farm outside of Heidelberg with his family and a host of wild animals. He is author of *Die Technik der Wildphotografie*.

Helen Rhode (190) lives in a log cabin on the shore of Alaska's Kenai Lake. Her photographs are reproduced in many national publications.

Carl Roessler (271), author of *The Underwater Wilderness: Life Around the Great Reefs*, has led almost 300 diving programs all over the world. He has won numerous awards in international film competitions.

Edward S. Ross (148), Curator of Entomology at the California Academy of Sciences, pioneered in candid insect photography in his *Insects Close Up* (1953). He has photographed nature in almost every part of the world.

Leonard Lee Rue III's (108, 136, 207) photographs have appeared in over 1000 publications in 32 countries. He has written 15 books and averages 200 lectures per year.

George Schaller (154, 155) is a zoologist with the New York Zoological Society. He has published numerous books based on his research in Africa, India, Nepal and Pakistan.

Hans Silvester (230) is a Frenchman who has done much wildlife photography and is particularly known for his book *The Horses of Camargue*.

Jim Simon (30) was one of the original Disney photographers and later a public information officer for outdoor tourism in Wyoming.

Alvin E. Staffan (82, 89, 90) is both nature artist and photographer for the Ohio Department of Natural Resources. His photographs appear in many publications.

Bob and Jill Stoecker (177, 210) are animal ecologists in Boulder, Colorado. They have photographed throughout the western United States and in East Africa.

Mark Stouffer (150, 178) and his brother Marty have combined documentary film making talents to produce several works, including an NBC-TV Network Special entitled *The Predators*.

Soames Summerhays (265) is a British marine biologist associated with the Barrier Reef Authority in Australia.

Charles G. Summers (108), now living in Littleton, Colorado, has had photographs of animals in the Time-Life animal series, in Chanticleer Press books, and *National Wildlife*.

Jan Thiede (241) has photographed wildlife throughout East Africa. A graduate of the Pratt Institute, she is an audiovisual designer.

Dale R. Thompson and George D. Dodge (121) have made wildlife films for the National Geographic Society and Warner Brothers. Their photographs have appeared in numerous magazines.

Jean-Phillipe Varin (37, 98, 278), a biologist-photographer, is co-author of *Photographing Wildlife*.

Albert Visage (37, 92, 98) was an electrical engineer before becoming a wildlife photographer. His work has been widely published.

Steven Wilson (198, 200) attempts through films (he has made five) and widely published photographs to "increase man's awareness of the other living things on our planet."

Michael Wotton (91) grew up in England but now lives in the Pacific Northwest and is associated with the Weyerhaeuser Company's Forestry Research Association.

Belinda Wright (2, 179, 180, 228, 238) lives in India. She has co-produced a series of wildlife films for Time-Life, and her pictures have appeared in many publications including *National Geographic*.

Jonathan Wright (138), a mountaineer from Aspen, Colorado, has had work reproduced in *Natural History*, *National Geographic* and *National Wildlife*.

Index

Numbers in italic indicate pictures

Aardvark, 114, 115
Acinonyx jubatus, 10, 143, *156, 158*
Acrobates pygmaeus, 60
Aepyceros melampus, 194
Ailuropoda malanoleuca, 164, *168*
Ailurus fulgens, 164, *169*
Alcelaphus buselaphus, 194
Alces alces, 174, *188-189*
Alouatta seniculus, 38
Alpaca, 218, 219, *221, 222*
Anoura geoffroii, 71
Anteater, 103, 114, 115
    giant, 115, *122*
    short-nosed spiny, *121*
    silky, 115
Antechinus, 54
Antelope, 127, 193, 194-195
    goat, 246
    nyala, *204*
    oryx, *206*
    pronghorn, 126, *200*
    sable, *208*
    saiga, 196
Antidorcas marsupialis, 202
Antilocapra americana, 200
Aplodontia rufa, 86, *91*
Arctocephalus australis
    galapagoensis, 265
Armadillo, 113
    giant, 115
    hairy, 115
    pichi, 115
    three-banded, 115
Ass, wild, 218
    Indian, *228-229*
Ateles, 23
Axis
    axis, 174, *180*
    porcinus, 176, *179*
Aye-aye, 23

Baboon, 24
    Chacma, *42-43*
    gelada, 24, *37*
Badger, 102
    American, *108*
    ferret, 103
    honey, 101, 102
    Malayan stink, 103
Balaena glacialis, 286
Balaenoptera musculus, 264
Bandicoot, mouse, 54
Barasingha, 174
Bassaricyon gabbii, 164
Bassariscus astutus, 162
Bat, 18, 69-72
    bamboo, 70
    big brown, 70
    bulldog, 70, *76-77*
    flower, 71
    fruit, 72
        Egyptian, *78*
    leaf-nosed, *74*

little brown, 70, *82*
long-nosed, 71, *80, 81*
nectar-feeding, 71
spear-nosed, 71
vampire, 71, *73*
Bear, 101, 161-164, 176
    American black, 161, *177*
    Asiatic black, 163
    brown, 161, 163, *165, 166*
    Eurasian brown, 163
    grizzly, 163, *165, 166, 177*
    Malayan sun, 163
    polar, 161, 162-163, *170*
    sloth, 163
    spectacled, 163
Beaver, 17, 87, 88
    mountain, 86, *91*
    North American, *95*
Birds, 15-18
Bison
    bison, 193, *197, 198*
    bonasus, 193
Bison, American, 193, *197, 198*
Blarina brevicauda, 116
Blastocerus dichotomus, 174
Boar, wild, 143, 217
Bobcat, 142, *150*
Bos
    gaurus, 195
    grunniens, 218
Bradypodidae, 113
Bradypus, 119
Bubalus bubalis, 218
Buffalo, 16-17, 143
    African, 102, 194, 195, *210*
    water, 218
Bushbuck, 193

Cacajao rubicundus, 37
Caenolestidae, 56
Caiman, 143
Callicebus, 34
Callithricidae, 23
Callithrix aurita, 30
Camargue, 217-218
Camel, 219, 262
    one-humped (dromedary), 219
    two-humped (Bactrian), 219,
    226-227
Camelus
    bactrianus, 219, *226-227*
    dromedarius, 219
Canidae, 125
Canis
    aureus, 126
    latrans, 126, *138*
    lupus, 126, *132*
Capra
    falconieri, 246
    hircus, 245
    ibex, 256
    ibex sibirica, 245
Capreolus capreolus, 174
Capybara, 86
Caracal, 142, *149*
Caribou, 143, 173, 176, *190*
Carrighar, Sally, 18

Castor, 87
    canadensis, 95
Cats, 141-144
    ringtail, 162, 164
Cattle, 262
    domestic, 17, 217, 218
    wild, 193, 217
Cebidae, 23
Cebuella pygmaea, 32
Ceratotherium simum, 234
Cercartetus, 54
Cercopithecidae, 23
Cercopithecus neglectus, 36
Cervus
    duvaucelli, 174
    elaphus, 175, *186*
    nippon, 175
    unicolor, 182
Cetaceans, 262, 264
Chaetophractus villosus, 115
Chamois, 246-247
Cheetah, *10,* 143-144, *156, 158*
Chicken, 102
Chimpanzee, 22, 23, 24, *47*
Chironectes minimus, 56
Chital, *see* Deer, axis
Clethrionomys rutilus, 91
Coati, 162, 163
Coendou, 120-121
Condylura cristata, 114
Connochaetes taurinus, 193, *212*
Cottontail, *see* Rabbit
Cougar, 17, 101, 142, 143, *149*
Coyote, 17, 125, 126, *138*
Cricetus cricetus, 87
Cuon alpinus, 126, *129*
Cuscus, 55
    spotted, *57*
Cyclopes didactylus, 115
Cynomys
    ludovicianus, 96
    parvidens, 86

Dama dama, 175
Dasypodidae, 113
Dasyurus, 55
    geoffroii, 62
Daubentonia madagascariensis, 23
Deer, 102, 126, 142, 173-176
    axis, 174, *180, 183*
    barking, 141
    Chinese water, 173
    fallow, 175
    hog, 176, *179*
    mule, 175, 176, *177, 178*
    musk, 173, *179*
    pampas, 174
    Père David's, 174
    red, 175
    roe, 174, 176
    sambar (Aristotle's), 125, 141, *182*
    sika, 175
    swamp, 174
    whitetail, 174, 176, *184*
Delphinus delphis, 282
Dendrolagus, 53
Desman, Russian, 114

Desmana moschata, 114
Desmodontidae, 71
Desmodus rotundus, 73
Dhole, 125, 126, 127, *129*
Diceros bicornis, 234
Didelphidae, 56
Didelphis virginiana, 56
Didermocerus sumatrensis, 234
Dinosaur, 21
Dipodomys, 87
Dog, 102, 103, 162, 163
   African hunting, 126-127, *130-131*
   German shepherd, 115
   wild, 125-128, 143
Dolphin, 17, 261, 262, *282*
   bottle-nose, 264
   river, 261
Donkey, 218
Dugong, 261, 262, 264
Dugong dugon, 264
Dunnart, 54

Echidna, 113, 114, 115, 116
Elaphurus davidianus, 174
Elephant
   African, 17, 233, 235-236, *240-241*
   Asiatic, 236
Elephas maximus, 236
Elk, 18, 102, 143, 175, 176, *186*
Enhydra lutris, 262, *281*
Eptesicus fuscus, 70
Equus
   asinus, 218
   burchelli, 196, *214*
   caballus, 217
   caballus przewalskii, 217
   hemionus khur, 228-229
Erethizon dorsatum, 116, *120*
Erinaceus europaeus, 114, *120-121*
Eumetopias jubata, 266
Euro, 55

Felis
   caracal, 142, *149*
   concolor, 142, *149*
   lybica, 142
   lynx, 143, *149*
   manul, 148
   pardalis, 148
   rufus, 142, *150*
   serval, 142, *148*
   wiedii, 141
Ferret, black-footed, 17
Fisher, North American, 102
Flamingo, greater, 217
Fox, 217
   gray, 125, *136*
   red, 125, *135*
Fruit eater, 70

Galagidae, 23
Galago, 23
Gaur, 143, 195
Gazella
   granti, 194, *203*
   thomsoni, 195
Gazelle, 142, 143, 193

Grant's, 194, *203*
   Thomson's, 195
Geomyidae, 86
Georychus capensis, 86
Gerbil, 87
Gibbon, 17-18, 21, 23
   white-handed, *49, 50*
Giraffa camelopardalis, 234, *242*
Giraffe, 234, *242*
Glaucomys, 85
Glider
   feathertail, *60*
   greater, 55, *58*
Goat, 245
   markhor, 246
   mountain, 246, 247, *258*
   Persian wild, 245, 247
Gopher, pocket, 86
Goral, 246
Gorilla, 21, 24, *48*
Gorilla gorilla, 21, *48*
Guanaco, 218, 219
Guenon, *36*
Gulo gulo, 101, *110*

Halichoerus grypus, 264
Hamster, 87
Hare, 86, 102
   Arctic, *97*
   mountain (blue), *98*
   snowshoe, 88, 143
   varying, 143
   See also Rabbit
Hartebeest, 194, 196
Hedgehog, 115, 116
   European, 114, *120-121*
Helarctos malayanus, 163
Herpestes edwardsi, 114
Hippopotamus, 236, *237*
Hippopotamus amphibius, 236, *237*
Hipposideros commersoni, 74
Hippotragus niger, 208
Horse, 143
   domestic, 217-218
   Mongolian (Przewalski's), 217, *230*
   wild, 217-218, *230*
Hydrochoerus, 86
Hydromys chrysogaster, 86
Hydropotes inermis, 173
Hydrurga leptonyx, 270
Hyena, 143
Hylobates, 21
   lar, 49, 50
Hypsiprymnodon moschatus, 55
Hystricidae, 116
Hystrix, 116
   africae-australis, 121

Ibex, *256*
   Nubian, 247
   Siberian, 245
Ichthyomys, 87
Ictonyx striatus, 103
Impala, 194-195, 196
Indridae, 23
Indris, 23
"Irish elk," 174

Jackal, golden, 126
Jackrabbit, 125
Jaguar, 143, *152*

Kangaroo, 55, 56
   great gray, 53, 55
   rat, 54-55
   red, 55, *66*
   tree, 53, 55
Kieran, John, 18
Kinkajou, 164
Koala, 54, *64*, 164
Kob, *207*
Kobus kob, 207
Kudu, greater, 194
Kultarr, 54

Lama
   glama, 218, *224*
   guanicoe, 218
   pacos, 218, *221, 222*
Langur
   golden, *36*
   Indian, *39, 40*
Lemming, 88
Lemmus, 88
Lemur, 23
   black, *25, 26*
   ring-tailed, *28*
Lemur
   catta, 28
   macaco, 25, 26
Lemuridae, 23
Leontideus rosalia, 31
Leopard, *5,* 143
   clouded, 142
   spotted, 142
Leporidae, 86
Leptonychotes weddelli, 263
Leptonycteris nivalis, 71, *80*
Lepus
   americanus, 88
   arcticus, 97
   timidus, 98
Lion, *6,* 142, 143, *154-155*
Llama, 218, 219, 224
Loris, 23
Lorsidae, 23
Loxodonta africana, 235, *240-241*
Lutra, 103
   felina, 103
Lycaon pictus, 126, *130-131*
Lynx, 142, *149*
   Canadian, 143

Macaca
   fuscata, 44-45
   mulata, 24
Macaque, Japanese, *44-45*
Macropodidae, 53
Macropus
   giganteus, 53
   robustus, 55
   rufus, 55, *66*
Manatee, 261, 262, *280*
Mandrill, 17, 21, *37*
Mandrillus sphinx, 21, *37*

Manis, 114
  gigantea, 115
  tricuspis, 118
Margay, 141
Marmoset
  golden lion, *31*
  pygmy, *32*
  white-eared, *30*
Marmot, hoary, 87
Marmota
  caligata, 87
  monax, 87
Marsupials, 53-56
Marten, 103, 104
  American, *108*
Martes, 103
  americana, *108*
  pennanti, *102*
Megaptera novaeangliae, *284*
Meles meles, 102
Mellivora capensis, 102
Melogale moschata, 103
Melursus ursinus, 163
Mephitis mephitis, *109*
Meriones, 87
Mice, *see* Mouse
Microperoryctes, 54
Microsorex hoyi, 116
Mink, 103, 104, *108*
Miopithecus talapoin, 23
Mirounga
  angustirostris, *275*
  leonina, *263*, *272*
Mole, 116
  eastern American, 114
  star-nosed, 114
Mongoose, Indian, 114
Monkey, 23, 143
  De Brazza's, *36*
  howler, *38*
  proboscis, *36*
  rhesus, 23
  spider, 23
  squirrel, 23
  talapoin, 23
  titi, *34*
Monotremes, 116
Moose, 101, 126, 143, 174, 175,
  176, *188-189*
Moschus moschiferus, 173, *179*
Mouflon, 245, 247
Mouse, 102
  deer, 87, *91*
  golden, *90*
  house, *89*
  kangaroo, *91*
Mundarda, 54
Muntjac, 141
Mus musculus, *89*
Muskrat, 85, *95*
Mustang, 218
Mustela, 102
  frenata, 102, *105*
  lutreola, 103
  ruxosa, 102
  vison, 103, *108*
Mustelids, 101-104

Mydaus javanensis, 103
Myospalax, 87
Myotis, 70
  lucifugus, *82*
Myrmecobius fasciatus, 54
Myrmecophaga tridactyla, 115, *122*
Myrmecophagidae, 114

Nasalis larvatus, *36*
Nasua, 162
"Native cat," 55
  western, *62*
Neofelis nebulosa, 142
Neomys fodiens, 116
Noctilio leporinus, 70, *76-77*
Notomys, *91*
Numbat, 54
Nyala, *204*

Ocelot, *148*
Ochotona princeps, *91*
Ochotonidae, 86
Ochrotomys nuttalli, *90*
Odobenus rosmarus, 264, *276*
Odocoileus
  hemionus, 175, *177*
  virginianus, 174, *184*
Olingo, 164
Ondatra zibethica, 85, *95*
Opossum, 18, 56
  Virginia, 56
Orangutan, 17, 23, *46*
Orcinus orca, 264
Oreamnos americanus, 246, *258*
Ornithorhynchus anatinus, 113, *117*
Orycteropus afer, 114
Oryctolagus cuniculus, 88
Oryx, 143, *206*
Oryx gazella, *206*
Ostrich, 193
Otter
  giant, 103
  marine, 103
  river, 103
Ovis
  ammon, 247
  canadensis, 245, *252*
  dalli, *249*
  musium, 245
Oxen, 218
Ozotoceras campestris, 174

Pachyderms, 233
Pallas' cat, *148*
Pan troglodytes, 22, *47*
Panda
  giant, 164, *168*
  lesser, 164, *169*
Pangolin, 114, 115
  African tree, 115, *118*
  giant, 115
Panthera
  leo, 6, 142, *154-155*
  onca, 143, *152*
  pardus, 5, 142
  tigris, 2, 141, *145*, *146*
Papio ursinus, *42-43*

Peromyscus maniculatus, 87, *91*
Petrogale, 55
Phalanger, 55
  maculatus, *57*
Phascolarctos cinereus, 54, *64*
Phoca
  groenlandicus, *270*
  vitulina, 262
Phyllostomus hastatus, 71
Physeter catodon, 263
Pika, 86, *91*
Pinnipeds, 162, 262, 264
Pipistrelle, 70
Pipistrellus, 70
Planigale (*Planigale*), 53
Platanistidae, 261
Platypus, 113, 114, 116, *117*
Peocilogale albinucha, 102
Polecat, 104
  marbled, 103
Pongidae, 23
Pongo pygmaeus, 23, *46*
Porcupine, 102, 115
  crested, 116
  North American, 116, *120*
  Old World, 116
  prehensile-tailed, *120-121*
  South African, *121*
Porpoise, 261, 262
Possum
  greater gliding, 55, *58*
  pygmy, 54
Potos flavus, 164
Potto, 23
Prairie dog, 17
  black-tailed, *96*
  Utah, 86
Presbytis
  entellus, *39*, *40*
  geei, *36*
Primates, 21-24
Priondontes giganteus, 115
Procyon
  cancrivorous, 162
  lotor, 161
Prosimians, 22-23
Pseudois nayaur, 245
Pteronura braziliensis, 103
Pteropus, 70
Pudu (*Pudu*), 174
Pusa sibirica, 261

Rabbit, 86, 88
  cottontail, 88, 101, *106-107*, 125
  See also Hare; Jackrabbit
Raccoon, 161-162, 163
  crab-eating, 162
Rangifer tarandus, 173, *190*
Rat, 143
  Australian thick-tailed, 86
  brown, 87
  Cape mole, 86
  cotton, 88
  fish-eating, 87
  kangaroo, 54, 87
  mole, 87
  naked-tailed, 86

water, 86
white-tailed, 87
"Rat opossum," 56
Rattus norvegicus, 87
Reindeer, 173, 176
Rhinoceros
    black, 234, 235
    Indian, 234, 236, *238*
    Javan, 234
    Sumatran, 234
    white (square-lipped), 234-235
Rhinoceros
    sondaicus, 234
    unicornis, 234, *238*
Ringtail cat, 162, 164
Rodents, 85-88, 102
Rousettus aegyptiacus, *78*
Rupicapra rupicapra, 246

Saguinus oedipus, *30*
Saiga tatarica, 196
Saimiri, 23
Sarcophilus harrisii, 55, *63*
Scalopus aquaticus, 114
Schoinobates volans, 55, *58*
Sciurus niger, *94*
Sea lion, 262, *270, 271*
    Galápagos, *268*
    Steller's, *266*
Sea mammals, 261-264
Sea otter, 17, 262, *281*
Seal, 261, 262, 263
    elephant, 263, *272, 274, 275*
    fur, 17, 18, 262, *265*
    gray, 264
    harbor, 262
    harp, *270*
    Lake Baikal, 261
    leopard, *270*
    Weddell, 263
Selenarctos thibelanus, 163
Serengeti Plain, 193, *212*
Serow, 246
Serval, 142, *148*
Sheep, 245, 262
    argali, 247
    bighorn, 245, 246, 247, *252, 254*
    blue, 245, 246
    Dall, *249, 250*
Shrew, 113, 116
    musk, 116
    pygmy, 116
    short-tailed, 116
    tree, 22
    water, 116
Siamang, 21
Sigmodon, 88
Sirenians, 262, 263, 264
Skunk, 103, 104
    hog-nosed, 101
    striped, *109*
Sloth
    three-toed, *119*
    tree, 113
Sminthopsis, 54
Soricidae, 113
Springbuck, *202*

Squid, 263
Squirrel
    African ground, *92*
    flying, 18, 85
    fox, *94*
Stag, giant, 174
Suncus etruscus, 116
Sylvilagus, 88
Symphalangus syndactylus, 21
Syncerus caffer, 194, *210*

Tachyglossidae, 113
Tachyglossus aculeatus, *121*
Takin, 246
Tamandua, 115
Tamandua tetradactyla, 115
Tamarin, cottonhead, *30*
Tarsier, 23
Tarsiidae, 23
Tasmanian devil, 55-56, *63*
Taxidae taxus, *108*
Tayra, 103
Tayra barbara, 103
Tenrec, 116
Tenrecidae, 116
Territoriality, 194-196
Theropithecus gelada, 24, *37*
Thylacine, 54, 56
Thylacinus cynocephalus, 54
Tiger, *2*, 17, 141, 142, 143, *145, 146*
Tolypeutes, 115
Tragelaphus
    angasi, *204*
    scriptus, 193
    strepsiceros, 194
Tremarctos ornatus, 163
Trichechus manatus, *280*
Tupaiidae, 22
Tursiops truncatus, 264
Tylonycteris pachypus, 70

Uakari, red, *37*
Urocyon cinereoargenteus, 125, *136*
Uromys caudimaculatus, 86
Ursus
    americanus, 161
    arctos, 161, *165, 166*
    maritimus, 161, *170*

Vampire, *73*
    false, 71
Vampyrum spectrum, 71
Vicuña, 219
Vole, northern red-backed, *91*
Vombatus ursinus, 54
Vormela peregusna, 103
Vulpes vulpes, 125, *135*
Wallaby, 56
    rock, 55
Walrus, 262, 264, *276, 277, 278*
Wapiti, *see* Elk
Weasel, 101-104
    African striped, 102
    least, 102, 103
    long-tailed, 102, *105, 106-107*
    winter white-coated (ermine), 104
Whale, 17, 261, 262, 263, 264

baleen, 264
blue, 264
humpback, *284, 285*
killer, 264
right, *286*
sperm, 263
toothed, 264
Wild cat, African, 142
Wildebeest, 18, 193, 196, *212*
Wisent, 193
Wolf, 17, 18, 126, 127, *132, 134*, 176
Wolverine, 101, *110*
Wombat, 54
Woodchuck, 87
Wuhl-wuhl, 54

Xerus erythropus, *92*

Yak, 218, 219
Yapok, 56

Zaedyus pichyi, 115
Zalophus californianus
    wollebaeki, *268*
Zebra, 196, *214*, 217
Zorille, African, 103
Zyzomys pedunculatus, 86